Groundwater Treatment Technology

Second Edition

Groundwater Treatment Technology

Second Edition

Evan K. Nyer

VNR VAN NOSTRAND REINHOLD
New York

Copyright © 1992 by Van Nostrand Reinhold

Library of Congress Catalog Card Number 91-42557
ISBN 0-442-00562-8

Manufactured in the United States of America

Published by Van Nostrand Reinhold
115 Fifth Avenue
New York, NY 10003

Chapman and Hall
2-6 Boundary Row
London, SE 1 8HN

Thomas Nelson Australia
102 Dodds Street
South Melbourne 3205
Victoria, Australia

Nelson Canada
1120 Birchmount Road
Scarborough, Ontario M1K 5G4, Canada

16 15 14 13 12 11 10 9 8 7 6 5 4 3 2

Library of Congress Cataloging-in-Publication Data
Nyer, Evan K.
    Groundwater treatment technology / by Evan K. Nyer.—2nd ed.
        p.   cm.
    Includes bibliographical references and index.
    ISBN 0-442-00562-8
    1. Water, Underground—Purification.   I. Title.
TD426.N94   1992
628'.3—dc20
                                                91-42557
                                                CIP

Contents

v

Preface

The purpose of the second edition of this book is to give a broad understanding of treatment technology for contaminated groundwater. The cleanup of groundwater is a unique problem. Although a variety of technology exists for the removal of compounds from water, the manner in which these techniques are applied as treatment systems will have to be altered for groundwater cleanups. This book will give the reader a general understanding of contaminated groundwater and specific knowledge in the application of existing groundwater cleanup technology.

Many treatment systems have been designed and installed since the first edition of this book was published. Data is available on existing technologies, and new technologies are quickly being applied to the field of groundwater treatment. Accordingly, I have expanded this book to include more details about each of the technologies, and new information about emerging technologies.

Some of the sections require a significant amount of detailed information, and I was forced to detour from the simple writing style of the first edition. The chapter on biological treatment required the most new information. However, I have tried to keep the detailed technical information in sections so that the readers may still read the chapter and get the general knowledge that they need, and then skip over any details that they do not need. The detailed technical information is then available to those readers that require the information.

Even with all of the knowledge that is now available, I still find that it is the basic information that is not being included in the designs. I have seen 3 million dollars spent on sample analysis at a site, and no one decided to spend 20 dollars on an iron analysis. I have seen multimillion dollar treatment system designs with no consideration given to life cycle changes in concentration or operator attention. While I have included all of the new information, I still emphasize the basics of groundwater treatment design. I have also tried to point out the practical parts of designing a treatment system.

This book will try to summarize the present knowledge and experience in the cleanup of groundwater. More importantly, the book will give engineers, scientists, and lawyers an understanding of how groundwater treatment systems are different from existing technology and of the methods that engineers must use to design a groundwater treatment system based on their present design knowledge.

The book will also provide, for the people who have general responsibility for the cleanup, a general knowledge in the different aspects of a cleanup. The goal of a cleanup cannot simply be to remove all of the contamination. The long-term solution must be permanent and economical. A "quick fix" may have problems two or three years after the cleanup. The most expensive method of cleanup may not be the best cleanup strategy for removing all of the contaminants. It is important for the design engineers and the managers who regulate the aquifer cleanups to understand what is possible and what is economical on a groundwater cleanup.

There is no way that an author can update a book like this and hold down a full time job. The only way that I was able to accomplish this task was to have some great help. Volunteers did significant research and provided first drafts for some of the sections. Others did reviews of sections to insure accuracy and readability.

In the general review area, I want to thank Richard Conway, David Miller, and Paula Magnuson. This is the second time for Richard, and his comments were once again very useful. David and Paula took time out of very busy schedules, and provided a nonengineering point of view. I also want to thank my secretary, Carla Gerstner. She did a lot of the typing, but more importantly, kept me organized during the project.

Each chapter had technical and writing help from various people. By chapter, I want to thank the following people:

Chapter 1—Jodi Montgomery
Chapter 2—Doug Mehan

Chapter 3—Robert Ackart, Frank Lenzo, Kevin Sullivan, Mark
          Stenzel, David Campbell, and Allison Vidal
Chapter 4—Gary Boettcher, Gwen Shofner, Jeff Greenwell
Chapter 5—Greg Rorech.

There is no way that the second edition of *Groundwater Treatment
Technology* would have been finished without these people. They all
used their free time in order to research the various technologies, and
to provide a first draft on that technology. The quality of the book
relies heavily on their assistance.

I also have to thank my employer, Geraghty and Miller, Inc. (G&M)
This company has provided me with the time and support that was
needed to finish this book. Computers, secretaries and drafting facilities were all made available. More importantly, G&M has provided an
amazing array of treatment projects for me to continue to develop my
expertise. They have also provided a work environment, and in my
opinion, the best group of professionals in the groundwater area today.

To provide accuracy and realism, I have used specific names of
products when discussing areas of technology. This does not mean
that either Van Nostrand Reinhold or I endorse these products. These
companies had knowledge that I hoped would be helpful to the
reader, and they were willing to share their knowledge with the public.
There are other companies in the field which provide the same
services. Several sources for specific technology and equipment should
always be contacted.

We have all learned a lot since the first edition of this book was
published seven years ago. For the most part, I am seeing this knowledge being applied to groundwater cleanups. I hope that this new
edition continues to increase the level of expertise available for this
important environmental area.

Groundwater Treatment Technology

Second Edition

1

Defining the Treatment System

During the past five decades, the environmental field has developed highly successful treatment techniques for the removal of contaminants from water. These methods have been developed in order to protect our national rivers, lakes, and other water bodies. Proven treatment methods are available for compounds ranging from human domestic waste to toxic organic waste and heavy metal contamination.

Within the last decade, we have become more aware, and are now taking action, to protect an additional body of water, groundwater. The main weapons in our fight against groundwater contamination have been the same methods that were developed for other water bodies. Almost all of the compounds that we are currently finding in groundwater have already been found in wastewater and a method for removing them has been successfully developed. In fact, multiple treatment methods are available for successful application. However, we must be careful of direct broad application of wastewater treatment methods to groundwater. While the basic treatment techniques will be the same, the method that we use to engineer and design these treatment methods will be different. The clean-up of groundwater is very different from the clean-up of wastewater. In addition, during the past few years, several new technologies have been developed specifically for groundwater and soil remediation. While these techniques have been devised for groundwater, we must still be careful to use the correct engineering and design methods when they are applied to remediation.

The most obvious difference between wastewater and groundwater management is that with groundwater remediations the body of water is actually being cleaned. In wastewater cleanups, we control and treat the wastewater that is entering a body of water. The body of water, a river or lake, actually cleans itself once we stop putting pollutants into it. Groundwater is not able to clean itself at a rapid rate. This is due to the fact that groundwater velocities and recharge rates of the required nutrients are much slower than in surface water. An additional difference is that the contaminants in groundwater must flow through a porous media (e.g., the rock or soil). Thus, in the clean-up of groundwater we must not only clean-up the source of the pollutants, but also the groundwater itself and the porous media through which the groundwater flows.

Other differences exist, and are used to define the parameters for treatment system design. Consider the black box approach to groundwater treatment system design, Figure 1-1. Some of the obvious parameters needed for design of groundwater remediations are: (1) flow; (2) influent concentrations; and (3) discharge requirements. The following sections will show how these parameters differ from wastewater treatment specifications and how to develop the data required for a groundwater treatment design. Groundwater remediation must also address the life of the project, including changes in

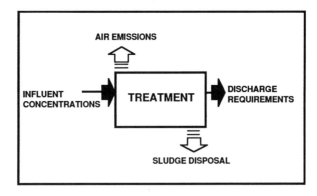

FIGURE 1-1. Parameters for defining a treatment system.

these parameters over time and maintain cost-effectiveness. Some of these factors are addressed in Chapter 2, Life-Cycle Design.

FLOW

A good place to start in determining the parameters that will define the treatment technique and final system design is flow. In wastewater treatment, the flow is a relatively simple parameter to determine. For domestic systems, the engineer selects a design date in the future. Most domestic systems are designed to last for 20 years. The engineer then uses population estimates for that time, multiplied by a standard factor for per capita water use. To this calculation, he adds flows from industrial waste in the city, and depending on the age of the sewer system, a flow factor for rain inflow and storm sewers.

Industrial wastewater flow is determined by adding up the expected flows from each of the unit operations at the plant. Once again, depending on the age of the plant, a factor is added for stormwater runoff. In both cases a safety factor is added, about 20%, to the final figure. For both cases, flow is not a design variable, it is a set parameter that must be determined by the engineer.

In the clean-up of groundwater, flow is a design variable. The engineer does not add up the different parts of the total flow. Instead he must weigh the effect of flow on the total cost of the system and on the time that is needed for a final clean-up. Let us look at the different factors that determine the flow for a groundwater treatment system.

Groundwater systems have many variables which impact flow and must be considered when designing a remediation system. These variables include the type of soil or rock (aquifer) through which the water must pass, the ability of the water to pass through the soil or rock (hydraulic conductivity), the type of contamination traveling through the ground, and the hydraulic gradient of the design area.

Figure 1-2 shows a typical spill situation that has entered an aquifer. Subsurface contaminant flow has two components, a vertical and a horizontal component. The contaminant travels through the unsaturated zone (no water present), and encounters the aquifer. Under most conditions, groundwater is constantly moving, although this movement is usually slow (typically 1-100 feet/year). To determine

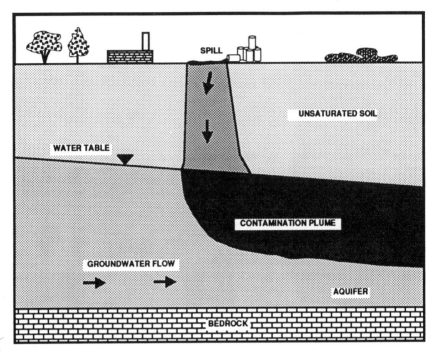

FIGURE 1-2. Contamination plume in an aquifer.

the flow and direction in an aquifer, basic information is needed. Once we collect or estimate that basic information, then the groundwater flow rate may be calculated. The relationship for flow is stated in Darcy's Law:

$$Q = -KA \, dh/dl$$

where:

Q = groundwater flow rate
A = cross sectional area of flow
dh/dl = hydraulic gradient: hydraulic head loss measured between two points (wells)
K = hydraulic conductivity, a measure of the ability of the porous media to transmit water.

To determine the direction and amount of flow, three or more wells may be drilled into the aquifer and the heads or water levels measured and compared to a datum (typically mean sea level). Groundwater will flow from high head to low head (the negative sign in Darcy's Law is for direction). The hydraulic conductivity (K) is a function of the porous medium (aquifer); finer grained sediments such as silts and clays have relatively low values of K, whereas sand and gravel will have higher values. Other physical factors may affect the hydraulic conductivity including porosity, packing, and sorting. The chemistry of the groundwater and the contamination may also affect K by causing contaminants to adhere to clay particles or by causing constituent precipitation (such as iron), both of which may reduce the permeability. As can be seen in Figure 1-2, the contamination plume almost always travels in the direction of the groundwater flow.

To move groundwater up to the treatment system, a well or trench may be constructed penetrating into the aquifer. A pump is then used to move the contaminated water to the surface. As was discussed earlier, groundwater flow is determined by the head differences measured in the aquifer. As we remove water from the aquifer by pumping, water levels and head relationships change. It can be seen by only adding one or two pumping wells to a groundwater remediation system that the overall groundwater flow patterns become very complicated. This is one reason why we need the assistance of hydrogeologists when designing a groundwater remediation system. Drawdown around a pumping well is called the cone of influence (Figure 1-3). The success of many groundwater remediation systems is dependent upon defining and understanding the dynamics of the cone or of the influence of the areas surrounding a pumping well.

Proper well placement and design in the appropriate hydrogeologic unit may stop and reverse the contamination plume. The well and pump will return the contamination above ground to the treatment system. The first factor in the groundwater recovery system design is the flow necessary to stop and reverse the movement of the contamination plume.

There are other methods which may be used to control the movement of the plume. These methods include the installation of hydraulic and physical barriers. Hydraulic barriers typically consist of numerous

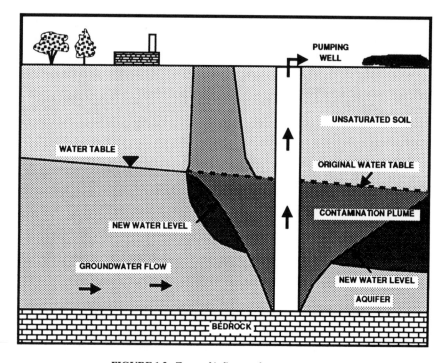

FIGURE 1-3. Zone of influence from a pumping well.

injection wells placed to alter the groundwater flow patterns. The hydraulic barriers may be placed downgradient of the contamination or surrounding the contamination. One of the great disadvantages of hydraulic barriers is that they may require a significant amount of clean water. Installation of injection wells may be limited by local, state, and federal permitting authorities or the actual physical properties of the soil or rock.

Physical barriers, such as clay slurry walls, sheet piles and grout curtains can be placed up or downgradient of the contamination or completely around the contamination plume. When correctly designed and constructed, a physical barrier can isolate the contamination. Eliminating water movement through the contaminated zone controls plume expansion.

Defining the physical setting for the subsurface barriers is necessary for ensuring their overall effectiveness. Project oversight by hydrogeologists, engineers familiar with underground structures, and geotechnical engineers is needed to determine the applicability of physical barriers. Barrier use may be severely limited by site characteristics and cost. These physical barriers must be keyed into confining layers so as to limit (horizontal/vertical) movement of contaminants. The deeper it is to a significant confining layer, the higher the construction expense. At depths greater than 60 to 100 feet, constructing a physical barrier of sufficient integrity is difficult and may require installation of an extensive monitoring system to review the barrier's effectiveness. Site characteristics alone may preclude the use of subsurface barriers. Regardless, the cost of their installation must be compared to the cost of stopping the plume movement by groundwater recovery. Even under ideal conditions, physical barriers will not be perfect and only slow the movement of the contaminant. On a practical level, if you think of physical barriers as a bank vault that will lock in the contaminant, you will be disappointed.

The aquifer pictured in Figures 1-2 and 1-3 depicts the "ideal" aquifer. In many cases, the bedrock or the geologic units below the base of the aquifer has a permeability high enough to act as a source of water through the upward leakage from the units below. Also, the thickness of the aquifer could be extensive and a plume may not reach the aquifer base. It is beyond the scope of this book to discuss the many groundwater and aquifer systems. The basic idea is that the contamination plume needs to be stopped and that one of the factors for controlling the plume is groundwater flow from the recovery well or trench. All of the design factors discussed in this chapter are discussed in basic terms. The "treatment design engineer" will always have to work with the people who are knowledgeable about the subsurface hydraulics. The reverse is also true. The final cleanup design should not solely rest with the hydrogeologists. The two disciplines must work together to develop the most cost effective remedial alternative.

The second factor that may have to be considered as part of the flow, is the amount of water entering the contamination area on a local or regional scale. One part of this has already been discussed. Slurry walls, or similar devices can interrupt the flow into or out of the contamination site. Two other possible entry points into the contaminated site are surface water from streams and lakes, or rain that can percolate down to the aquifer. In addition, the bedrock and other sediments below the aquifer may have higher permeability and may provide a source of groundwater by upward leakage.

Water from the surface can be controlled by either capping the contamination site with an impermeable layer or by providing good surface area drainage directed away from the site. Water coming from the bedrock is much harder to control and will probably have to be added to the flow for the treatment system.

The third factor in the flow to the treatment system is the speed with which the clean-up is to occur. In most groundwater systems, the more water that is pumped from the well, the lower the level of water surrounding the well. This will increase the head differential in the groundwater and force the water at a faster rate to the well. There is a limit as to how fast the water can travel through an aquifer and it is very easy, in many aquifers, to pump a well dry. The relationship between pumping rate, system design, and the time necessary to remediate a site is often complicated. The engineer or project manager must decide between the cost of increasing the treatment system size and the savings to be achieved by reducing the time for groundwater clean-up.

The relationship between system flow and clean-up time is complicated. When assessing any groundwater contamination site, it is necessary to identify the source or sources of contamination. The second step is to stop those sources from contributing to the already expanding contamination plume. Sources of contamination may be an underground storage tank, a surface source, or even residual contamination or product within the unsaturated portion of the subsurface. For example, Figure 1-2 shows the contaminants flowing through the unsaturated zone to the aquifer. Let us assume that a well is placed within the contamination plume and that maximum flow over a period of three months is sufficient to completely eliminate the plume. The equipment is packed-up and everyone leaves. A problem could arise if the source has not been identified and stopped. There are still contaminants in the unsaturated zone. These contaminants may still migrate to the aquifer and three months later another contamination plume could form.

One solution to this problem is that the original clean-up must be slowed so that the natural flushing methods have a chance to bring most of the contaminants through the unsaturated zone. Of course, an alternative would be to speed-up the natural flushing action (see

In-Situ Treatment in Chapter 4), and maintain the original speed of the clean-up.

Another example, one that cannot be solved as readily, is seasonal variation in the groundwater level. As one would expect, during times of high rain flow, spring and fall, the overall level in the aquifer can increase. The problem arises when the level recedes and the contaminants are left behind in the unsaturated zone. The next time that the water level is high or rain percolates through the ground, the contaminants will reenter the aquifer. In these cases, the best solution may be to slow down the flow to the treatment system and perform the clean-up over several high water seasons or to excavate and remove the contaminated soil.

Other factors will have a relatively minor effect on the flow of the treatment system. The number of recovery wells or trenches used will have an effect on the amount of flow for clean-up. The effectiveness of wells and trenches to pull in the contamination and to remediate the site is dependent on the location of the recovery system. Systems located on the edges of the plume may be successful in capturing contaminants located there but the overall success will be diminished because clean water from outside the plume will also be recovered. Successful recovery system locations also have to take into account "non-flow" zones caused by the design of the system. Non-flow or dead zones may successfully capture the contaminant by alternating the pumping schedule by turning some pumps on or off. A recovery system located closer to the leading edge of the plume will require a lower flow rate to stop the movement of the plume than will a well located at the site of the original contamination. Although the idea is to clean up the groundwater, many other factors may come into play when selecting recovery system locations. For instance, a second recovery system located near the center of the contamination may also be required. The larger the plume, the more likely the need for a second or multiple recovery system.

Discharge requirements may affect the flow. More flow will be required if some of the water is to be used to flush the unsaturated zone of contaminant. Water returned to the site can decrease the time for clean-up by increasing the hydraulic head and forcing the groundwater to the central recovery point at a faster rate. This may be limited by the hydraulic properties of the sediments above the groundwater. This same water will increase the total water entering the site and the flow to the treatment system will increase accordingly. The effect of discharge requirements will be discussed further in the third section of this chapter.

In summary, hydraulic properties of the aquifer and sediments above the aquifer will control groundwater flow to the treatment system. The following factors have to be considered when determining the flow to the groundwater treatment system:

1. Stop/reverse movement of the contamination plume,
2. Amount of water entering the contamination site on a local level,
3. The rate at which the clean-up is to occur,
4. The recovery system to be used, and
5. The final disposal/use of the treated water.

INFLUENT CONCENTRATION

The concentration of contaminants in groundwater are normally determined by sampling the water from a well. The types of contaminants will depend on the material originally lost to the ground. The relative concentration will depend on where the well intersects the contamination plume. The farther away from the original spill site, the more dispersion and contaminant degradation will occur to the original contamination plume. Several wells will have to be constructed to get a full picture of the plume (and the hydrogeology). The hydrogeologist may provide insight and guidance in the selection of wells for the monitoring system.

In addition to the distance from the spill site, several other factors affect the concentration of the contaminants. These factors include: amount of contaminant reaching the aquifer, solubility and density, amount of groundwater flow, distance to the water table, and time. These factors will all affect the size of the treatment system and the length of time that the system must be run for a full clean-up.

The amount of contaminant reaching the aquifer is made up of several components. The first is the amount of contaminant lost to the ground. The first question to ask is whether the source of contamination has been shut off. There are certain cases in which the contamination is still being introduced into the ground. An example of this is a landfill. Other cases also exist and may include hydrocarbons trapped in the unsaturated zone of a gas station. Leachate generation is also an example of a continual contaminant source. Leachate is generated when water flow through the contaminated zone causes a continual addition of new contamination to the groundwater (subsurface).

Once the source of contamination is defined and shut off, the hydrogeologist may assist in attempting to determine the total amount of material lost to the ground. Not all of the material lost to the ground will reach the aquifer. The unsaturated zone above the aquifer will adsorb a percentage of the contaminants. It is widely reported that only 50% of the gasoline in a spill from a storage tank normally reaches the aquifer. Of course, this amount depends on the type of soil in the unsaturated zone, the distance to the aquifer, and the total amount of material spilled. The material left in the unsaturated zone will be discussed further in the next section of this chapter and in the In-Situ Treatment section of Chapter 4.

The next component of contaminant concentration in the aquifer comes from the solubility and the density of the material spilled. A large portion of the materials spilled into the ground is not soluble in water. When the material not soluble in water reaches the aquifer, it does not mix. Figure 1-4 shows a gasoline spill. The main component of the gasoline does not enter the aquifer. It floats on top of the aquifer and spreads in all directions. A small percentage of the gasoline, mainly benzene, toluene, xylene, and ethyl benzene (BTXE) compounds, typically found in unleaded gas, does enter the aquifer and they may form the normal plume.

Gasoline is lighter than water, so it normally floats on top of the aquifer. Most straight chain hydrocarbons are lighter than water and will stay on top of the aquifer. Compounds that are heavier than water usually sink to the bottom of the aquifer. Figure 1-5 shows a trichloroethylene spill. Chapter 3 provides the densities of 30 compounds. This list, combined with the solubility of the compound, will help assist the groundwater scientist in determining where to find a particular compound once it is spilled into the ground. The section on Pure Compound Recovery in Chapter 3 discusses locating and removing these compounds from the surface.

The amount of water flow through the aquifer will also have an affect on the contaminant concentration in the aquifer. The greater

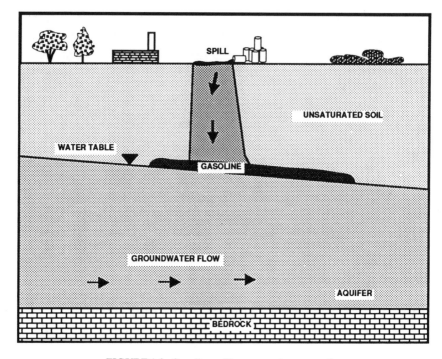

FIGURE 1-4. Gasoline spill encountering an aquifer.

the amount of water that passes the original spill site, the greater the dilution or dispersion that will occur. Once again, the groundwater scientist must decide on the economics of limiting the groundwater flow with a physical structure or by hydraulic controls.

Finally, these components combine to form a time effect on constituent concentrations. There are three patterns that constituent concentrations follow over the life of the project. These patterns are summarized in Figure 1-6. First, there is the constant concentration exhibited by a leachate. If we do not remove the source of contamination, then the source will replace the contaminants as fast as they can be removed with the groundwater system. Until the source of contamination is remediated, the concentration will remain the same. We normally think of "mine" leachate or "landfill" leachate. But, anytime there is a continual source of contamination, we are dealing with a leachate.

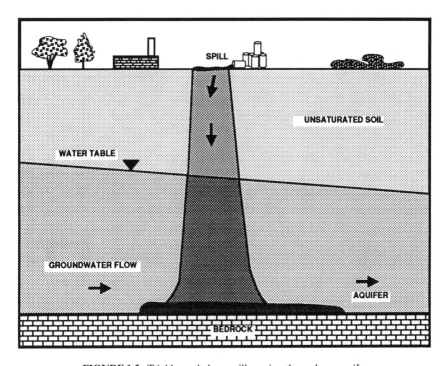

FIGURE 1-5. Trichloroethylene spill passing through an aquifer.

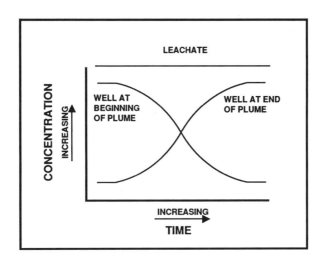

FIGURE 1-6. Time effect on concentration found in a well.

The second possible pattern arises when the contamination plume is being drawn toward the groundwater removal system. This mainly happens with municipal drinking water. In this situation, the concentration increases over time. The well is originally clean, but becomes more contaminated as the plume is drawn toward the well.

The final pattern is associated with remediation. In this case, if the original source of contamination is removed, the concentration of the contaminants decreases over time. This is a result of retardation, natural chemical and biochemical reactions, and dilution of the surrounding groundwater.

Concentration data may be collected by drilling wells and by sampling the groundwater in the wells. The separate data from the wells can be combined in different ways. One can average all of the data, use a weighted average reflecting the influence of specific wells, or just take the maximum concentration found in all of the wells. The weighted average is probably the most accurate method and the hydrogeologist may assist by providing the weight criteria. The maximum concentration is the least accurate method. However, the maximum concentration method is being widely used. I guess the logic is that if the treatment system can handle the high concentration, there will be no problems with lower concentrations. This is also the method that is used for wastewater. But, this is a dangerous design method for groundwater. Several treatment methods will have a catastrophic failure if the concentration goes below a minimum (biological and heavy metal removal are two examples). Other treatment methods will not be optimized based upon maximum concentration.

Accurate concentrations are required for proper design of treatment systems. My experience has shown that the weighted average (the more accurate the weight factors the more accurate the final concentration) is the best method to employ.

One practical suggestion to close this concentration discussion. Most sites require an aquifer pumping test as part of the site studies. The aquifer pumping test can also be used as a method to gather concentration data. Concentration samples should be collected near the end of the aquifer pumping test. If there is a desire to see the immediate time effect on concentration, then use a general organic measurement (total organic carbon; total petroleum hydrocarbons, etc.) on samples taken every half to full hour. A full analysis can be completed two to three times during the testing period.

DISCHARGE REQUIREMENTS

The treatment plant design will also depend on the final disposition of the treated water. The engineer must decide whether the water will be discharged to a surface water body, to another treatment system, to a direct use, or returned to the ground or aquifer. The discharge requirements for each of these cases will have a major effect on the size and complexity of the treatment system.

The discharge may be sent directly to a stream or other surface water body. Such discharges are regulated by the National Pollutant Discharge Elimination System (NPDES) program. It must be determined if the discharge will be direct or indirect, (discharges to publicly owned treatment works (POTWs) are indirect), if it is a regulated categorical industry discharge, or if it requires bioassay testing prior to discharge. The following would be typical effluent requirements for a direct discharge: 30 mg/l (milligrams/liter) biochemical oxygen demand (BOD), 30 mg/l total suspended solids (TSS), 5 mg/l oil and grease, less than 1 mg/l for any heavy metal, and between 0.05 and 0.5 mg/l for specific toxic organics. Required monitoring and reporting associated with permitted discharges may add significant costs to the treatment system. The required discharge limitations may be stringent if the receiving body lies in a sensitive environmental setting.

Typically, the preferred method of final disposal for recovered groundwater is discharge to another treatment system, either a POTW, or an industrial wastewater treatment system. One of the advantages of discharging to the POTW is that the groundwater can be discharged to nearby sewer lines. The advantage of discharging to the industrial wastewater system, is that the type and concentration of the contaminants can be higher than would normally be allowed for an NPDES discharge.

The contaminants will have to be compatible with the waste that is already processed by the off-site treatment system. Some pretreatment may be necessary for either type of plant. The industrial plant may require neutralization, heavy metal removal or pure compound removal. The POTW will usually require that the water entering the sewer not exceed the normal concentration of domestic waste. The following would be typical limits for discharges to a POTW: 250 mg/l (BOD), 250 mg/l TSS, 100 mg/l oil and grease, less than 1 mg/l for any heavy metal, and between 0.5 and 5 mg/l for specific, toxic organics.

Regulations for each treatment plant will be different, and the potential plant will have to be contacted and the discharge limitations established. Local ordinances also govern discharges to POTWs. It is also far-sighted to review the POTWs compliance history with their NPDES permits. It is not recommended to discharge to a POTW that may be under review for significant noncompliance.

For an increasing number of situations, spills have affected a drinking water source or other final use source. To reuse the treated groundwater in these cases requires treatment using stringent effluent discharge requirements. However, the contaminant concentration is normally very low or the well would have been abandoned. Federal drinking water standards should be reviewed to determine discharge requirements.

Finally, the water from the well can be returned to the ground. If the water is to be returned to the ground, the recharge system should be strategically placed to affect the movement of the plume. Figure 1-7 shows the water being returned at the end of the plume. This will

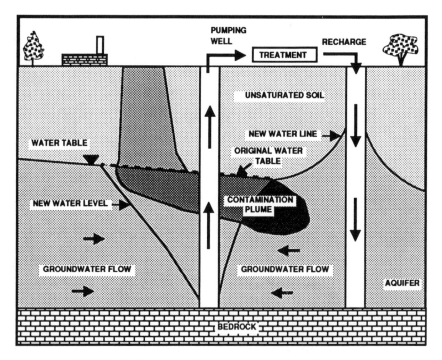

FIGURE 1-7.   Effect of a recharge well on groundwater flow patterns.

increase the hydraulic head, and force the plume back to the central well.

The recharged water may also act to flush the unsaturated zone of contaminants. Figure 1-8 shows the treated water being returned at the surface where the spill originally occurred. This process used to maximum capabilities is referred to as In-Situ Treatment. In-Situ Treatment will be discussed in Chapter 4. If the water is to be used at the end of the plume, the discharge requirements will be strict. This is because not all of the water will return to the central well. Any water escaping the zone of influence should be at background aquifer concentrations.

If the water is to be used to flush the contaminants or as part of an In-Situ Treatment, the level of treatment will be much lower. In fact, in the case of biological treatment, leaving some of the bacteria and the resultant enzymes will help solubilize, flush, and degrade the contaminants in the soil.

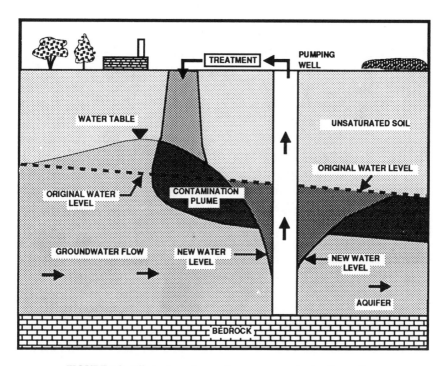

FIGURE 1-8. Effect of a surface recharge on groundwater flow patterns.

GATHERING A COMPLETE SET OF DATA
FOR THE TREATMENT DESIGN

We have now reviewed the major components needed to design a groundwater treatment system. Even with all of this data, we cannot be completely certain that the design will be perfect for the groundwater that we pump at the start of the project. One of the continuing problems with groundwater is determining how much data is needed to accurately define the actual conditions. All we really have is numbers on a piece of paper. Do these numbers represent what the groundwater treatment system will see when it starts up? How many data points are needed to define a concentration? How many data points are needed to define a flow rate?

It is very easy to turn a site into a "pin cushion" and install numerous borings and wells. It is also easy to take a multitude of samples on each well and spend millions of dollars on analysis. Even after all of this effort and money, there will still be a significant amount of uncertainty with the design data.

This is the result of the highly variable physical conditions in the subsurface. It is impossible to obtain detailed physical information on the unsaturated zone or the aquifer. Geotechnical engineers have used an "observational" design method originally developed by Karl Terzaghi and R. B. Peck[1] to develop designs for foundations, dams, etc., for many years. This method uses limited data from the site, incorporates experience from similar situations, and finally produces a design that can be slightly modified as the last of the data arrives during construction or start up.

The observation method of design also has a place in the design of groundwater treatment systems. We will never be able to gather enough data to be 100% certain of the design. We must use a combination of data from the field and experience from similar installations. In addition, our final design for the treatment system should include the ability to adjust to further changes as we gain full understanding of the nature of the groundwater from the actual pumping.

COMPLETING A REMEDIATION PROJECT

Most remediation plans include a final concentration that the project is trying to achieve. However, we do not analyze what these concentrations mean to the design of the treatment system and the operation of the project in general. We must understand what we mean by the end of the project, and we must set up a strategy to reach that finish point.

Life Cycle of a Project

Figure 1-9 shows the normal life cycle concentration of a remediation project. This is the same as the "well at the center of the plume" curve found in Figure 1-6. We have already discussed the beginning portion of the curve. The main point being that we cannot simply use the concentration that we find during the remedial investigation as the only criterion to design a groundwater treatment system. The concentration changes over the life of the project, and any design must address all of the concentrations encountered during the entire project. We also have to consider the bottom part of the curve. As the project progresses, the rate of removal decreases. As you can see in Figure 1-9, the curve tends to become parallel to the horizontal axis. There are several factors that contribute to the flattening of this curve, and they will be discussed later in the book. For now, let us build a framework to help us understand how to reach the end of the project.

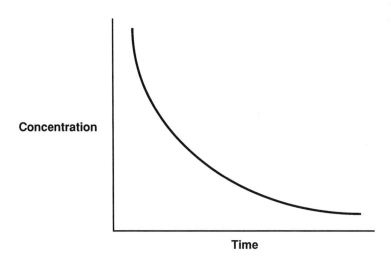

FIGURE 1-9. Life cycle concentration of a remediation project.

What Is Clean

The objective of most remediations is to clean the site. This raises the questions what is clean, and how do we define clean? The three main, conservative methods used to establish clean are: risk assessment; federal drinking water standards; and analytical detection limits. All of these methods have advantages and disadvantages.

Risk assessment can be uniquely designed for the specific site. Any unusual paths for human contact, and any design method used to prevent human contact can be incorporated into the risk assessment. However, there are no official standard methods to develop the risk formulations. Different basic assumptions will result in different specific numbers calculated for clean. Some states will not allow risk assessment to be used because of the possible variable results from the same data. While risk assessment may be the only method that can consider local anomalies, the variable output may cause long discussions on the reality of the numbers.

The federal drinking water standards are another source of numbers that can be used to establish what is clean. These numbers are set and specific. Their bases were published and discussed before the final figures were made official. Over time, more compounds will be included on the list. The only problem for remediation sites is to establish the relationship between the soil and the groundwater. All of the drinking water standards are based upon concentrations in water. The organics adsorbed to the soil particles in the vadose zone and the aquifer will slowly be released into the groundwater. While drinking water standards have a strong technical basis, this method does not directly address the adsorbed material.

Detection limits are the third method. For compounds that are highly toxic or in situations in which the numbers developed during a risk assessment are less than the detection limit of the compound, then the analytical detection limit can be used to establish what is clean. The main problem with this method is that the ability to detect a compound continuously improves. No one should accept "detection limits" as clean. Instead, the detection limits should be used as a basis for a specific concentration. Without a specific number, what is clean will change over the life of the project.

Any of these methods can be used to determine what is clean. However, if we look at the basis for each of these methods, we are

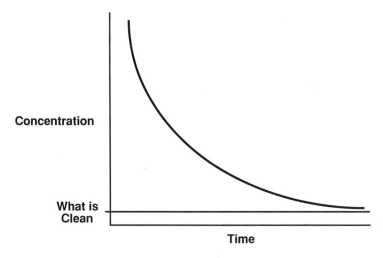

FIGURE 1-10. Achieving "clean" during a remediation project.

basing these concentrations on contact with human beings. In other words, clean is when the groundwater or soil is safe for human consumption. The end of the remediation occurs when the site is safe for people.

The problem with this definition of clean is shown in Figure 1-10. As the site gets closer to clean, the contaminant concentration reaches its asymptote. Figure 1-10 represents a worst case scenario in which the site never reaches clean. Even in cases where the site does reach clean, it will take many years.

During the last years of the project, the treatment system suffers from diminishing returns. The treatment system continues to run, but the amount of material removed is minimal. The money is still being spent on the site, but there is little return for this financial outlay.

Retardation vs. Biochemical Activity

A second factor comes into consideration when we approach the asymptote of the life cycle concentration. Natural biochemical reactions may be occurring at the same rate as the natural release of the contaminants due to retardation factors.

As we will discuss in Chapter 4, natural bacteria exist throughout the soil, vadose zone, and aquifer. If no toxic conditions exist, then the natural bacteria are already degrading the contaminants. Their rate of degradation is limited by the presence of oxygen and nutrients. Oxygen is usually the main limiting factor. In fact, biodegradation models, like BIOPLUME II[2], use oxygen concentration as the limiting factor in the degradation rate.

The natural rate of degradation is small compared to an above ground reactor or to enhanced in situ biological reactions. The aquifer or vadose zone have a limited capability to replenish oxygen used by the bacteria. But, when the contaminant concentration reaches very low levels, then the aquifer and vadose zone can naturally supply what is needed. At the same time, all of the contaminants do not release to the water at the same rate. The contaminants are adsorbed to the soil. Depending on the organic content of the soil and the chemical properties of the contaminant, the individual compounds will release to the water at a different rate. In addition, some of the contaminants will have to diffuse from their location to a place where they will be part of the main flow of the aquifer. This water will then flow out of the aquifer and into the treatment system.

Other researchers[2,3] have shown that a plume of degradable organics would not move through an aerobic section of an aquifer. Data was from actual contaminated sites. Both projects showed that pumping would not improve the remediation. In fact, more compounds would be exposed to humans from a pumping system than by allowing the natural bacteria to degrade the compounds in situ. Both articles showed that natural degradation could be at the same rate as plume migration. We can project a complete natural elimination of the plume based upon the same data.

At some point, the rate at which the compounds can be removed from the aquifer by pumping will be the same as occurs from the natural biological degradation rate of the aquifer. At this point, we will not be able to speed the cleanup no matter how fast we pump the well. In fact, if the aquifer can naturally replenish the oxygen demand from the contaminants, then pumping will not increase the rate that we reach clean at all. Once we reach this point, then all of the money that we are spending on pumping and treating this water is going to waste. The pumps could be turned off and all of the equipment removed, and we would still reach clean at the time we would have by leaving the system running.

Active Management

We need to develop another point in the life cycle of the project that determines when we can turn off the treatment system. This is the point at which we can stop spending most of the money. Clean occurs when we can reuse the property. Active management occurs when we can stop spending money.

Figure 1-11 shows the life cycle curve with the clean line and an active management line. The active management line represents the point in the life cycle where pumping will no longer speed the cleanup of the site. We should stop spending money at this point. The clean line represents when we can use the water, and the site for human consumption.

This is very similar to the situations that occurred in the 1970s when we cleaned up rivers and lakes. We installed wastewater treatment systems on municipal and industrial wastewater. This stopped the contaminants from entering the water body (lake or river). The river or lake then cleaned itself. This took time, and we did not use the water body the day after we installed the treatment system. We waited until the water was clean, or safe to use.

Rivers and lakes can remediate themselves faster than can groundwater. One reason for this is that oxygen can transfer into these water

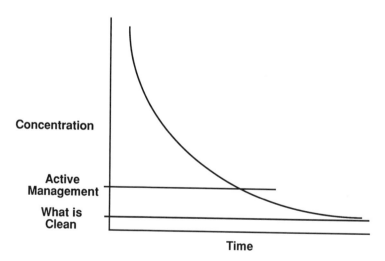

FIGURE 1-11. Active management vs. what is clean during a remediation project.

bodies faster than it can into groundwater. But, groundwater can remediate itself. The problem is that the rate is so slow that we normally find the time frame unacceptable. However, if we remove most of the contaminants, then the aquifer can finish the job on its own. If this rate is the same as the rate from pumping and treating, then there is no reason to continue the pumping and treating. We are at the point in the life cycle of the project when we must wait before we can use the water. Active management of the site should stop, and we should only monitor the site while we wait for clean.

Bio Modeling

The only problem left is to determine when we reach the active management point of the life cycle of the project. This point is basically a comparison of the normal fate and transport of chemicals to the biochemical reaction rate of those same chemicals. Therefore, some type of biomodelling will be required in conjunction with the solute transport materials.

We will discuss the details of the active management point throughout the book. At this point, it is only important to understand that the end of the remediation and the end of the groundwater treatment system are two separate points in a project.

References
1. Terzaghi, Karl and Peck, R. B., *GroundWater and Soil Contamination Remediation: Toward Compatible Science, Policy, and Public Perception.* National Academy Press, 1990.
2. Rifai, H. et. al. Biodegradation modeling at aviation fuel spill site. *Journal of Environmental Engineering* 114(5) October 1988.
3. Nyer, Evan K. Biochemical effects on contaminant fate and transport *Ground Water Monitoring Review* Spring 1991.

2

Life-Cycle Design

As we discussed in Chapter 1, most of the technologies used for groundwater treatment were developed for the wastewater market. We have already reviewed the differences between gathering the design parameters for the two types of systems. In this chapter we will discuss the changes of the equipment design itself caused by the unique situations we find in groundwater treatment.

Groundwater treatment is unique. In other types of treatment, we address the source of contamination only. The main goal of wastewater treatment is to stop or limit the amount of waste entering the body of water, e.g. river, lake, or ocean. Once we stop putting the pollution into the water, then the body of water actually cleans itself. With groundwater treatment, we are cleaning the body of water. As before, we must first remove all of the sources of contamination. But, we must continue and remediate the body of water, the groundwater. The groundwater actually does clean itself once the sources of contamination are removed. We just find that the natural rate of 300 to 30,000 years is a bit too slow.

A second area in which groundwater remediation design is unique, is life expectancy of the project. When we deal with wastewater treatment systems we normally design for a twenty to thirty year life. This number represents the life expectancy of the equipment not the life expectancy of the project. Except for leachate treatment, groundwater treatment systems will last for much shorter periods of time and the controlling parameter is usually the life expectancy of the project.

In addition to the short life of the project, the treatment systems will be relatively small. So small, that the design engineer can not merely use the designs previously developed for the wastewater market and simply shrink them. Other differences exist and we will refer to all of these by describing the design method for a groundwater treatment system as *Life-Cycle Design*. The most important factor in the correct life-cycle design is to take into account the time effect on the influent parameters.

TIME EFFECT ON INFLUENT PARAMETERS

Flow

The amount of water discharged from the treatment system will probably remain the same during the life of the project. The amount of water that is being reused on site may vary during the clean-up. This would include the periods when the water was being used to increase the hydraulic head at the end of the plume, when the water was being used to flush contaminants from the unsaturated zone, and when the water was being used for in situ treatment. For these three cases, the flow to the treatment system may change during the project.

Water may be reused to increase the hydraulic head and remediate the plume in cases when the end of the plume is relatively far from the central well or wells and in cases when additional time for clean-up is not available. Near the end of the remediation period, the plume will be forced back to the original contamination site. There will no longer be any reason to reuse treated water to help move the plume. However, injection water may still be used to speed the progress of the project.

The same thing may happen when water is being reused to flush the contaminants from the unsaturated zone. Near the end of the project, the contaminants will have been flushed and only aquifer clean-up may be necessary. At this point, the engineer may decide to eliminate the reuse of the groundwater. This is also the case for in situ treatment. At the end of the project, the concentration of contaminants in the aquifer may be low enough to stop the recirculation of water. As in all good water treatment designs, provisions must be made for flow fluctuation. In addition to these major flow considerations discussed, other factors can affect the flow during the life of the project. Two main factors in groundwater treatment are drought and loss of power.

There are natural fluctuations in the level of groundwater during the year. Most designs are based upon the maximum water that the aquifer can transfer. However, during low water situations, as in droughts, water flow may be limited. The design should make provisions for low flow rates.

A second "natural" factor that can occur once the treatment system has been installed is power loss. Effects can range from frozen pipes, to overflowing tanks arising from loss of pumps, to the killing of bacteria in a biological system. While power loss is not normally a factor in choosing a particular treatment method, it is a practical factor that can cause any design to fail.

In all of these cases, the treatment plan engineer must take into consideration that the flow to the treatment system may change. The design engineer must make sure that the treatment system will function at all of the possible flow rates.

Concentration

In Chapter 1 we discussed the effect of time on concentration. Removal of the contaminants by the treatment system combined with the dilution effect by clean water entering the affected site produce a steady decline in the contaminants concentration in the groundwater. Figure 1-5 shows the decline of concentration over time. The design engineer can not assume that the lower the concentration the better the treatment system will operate. The design of certain processes are based on a minimum concentration. These units will loose efficiency with lower concentrations, and not function at all once a minimum concentration is reached. The design engineer must also take into account that the operational costs may be reduced as the concentration decreases. In either case, the design must be able to accommodate the entire life-cycle concentration of the project. Three examples of the effect of concentration on unit operations follow.

Treatment methods for organic contaminants are discussed in Chapters 3 and 4, and treatment methods for inorganic contaminants are discussed in Chapter 5. The reader should refer to these three chapters for details on the following examples. First, let us look at the effect of time on an organic contaminant that we will treat with biological methods, specifically using the activated sludge method.

Assume Figure 2-1 represents the influent life-cycle concentration. The flow(Q) will be 25,000 gal/day for the entire life of the project.

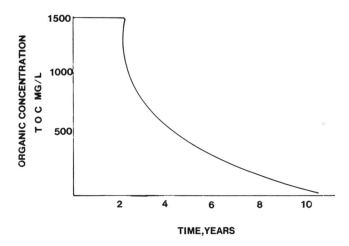

FIGURE 2-1. Life cycle concentration from a well at the center of the plume for an organic contaminant.

Also assume that all other environmental parameters are acceptable for biological treatment. Figure 2-2 pictures the proposed treatment system. As will be discussed in Chapter 4, the activated sludge process depends on the settling properties of the bacteria that are growing. To keep the bacteria in a growth phase in which they settle properly, the bacteria should have a sludge age (A) between 5 and 20 days. Let us look at what happens to the sludge age during the life of the project.

$$A = (X \times V)/(Q \times S \times Y)$$

Assume:

mixed liquor suspended solids, MLSS(X) = 3,000 mg/l
yield coefficient (Y)   = 0.25 lb/lb
volume of the aeration tank (V)   = 40,000 gal

for year 1, S = 1500 mg/l
    A = 12.8 days

for year 3, S = 1200 mg/l
    A = 16 days

for year 5, S = 600 mg/l
    A = 32 days

and for year 7, S = 300 mg/l
    A = 64 days

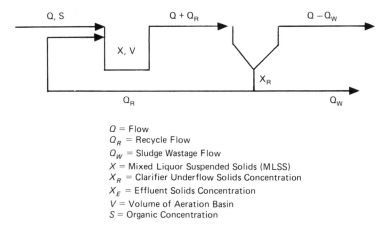

Q = Flow
Q_R = Recycle Flow
Q_W = Sludge Wastage Flow
X = Mixed Liquor Suspended Solids (MLSS)
X_R = Clarifier Underflow Solids Concentration
X_E = Effluent Solids Concentration
V = Volume of Aeration Basin
S = Organic Concentration

FIGURE 2-2. Activated sludge treatment system.

As can be seen from this data, the system will maintain the proper sludge age for about four years. After this time the sludge age will be too high, the bacteria will loose their settling properties, and the clarifier will not be able to separate the bacteria from the treated water. Once the clarifier fails, the system will not be able to maintain a high concentration of bacteria in the aeration basin. At this point, the system will no longer remove a high percentage of the incoming organic contaminants.

One solution to this problem is to lower the MLSS concentration. Figure 2-3 summarizes the sludge ages for the treatment system at MLSS level of 3,000 mg/l and 1,500 mg/l. This does extend the useful life of the treatment system, but the system still fails before the cleanup can be completed. And, there is a lower limit to the MLSS. The MLSS concentration entering the clarifier must be about 1,250 mg/l or above to ensure proper settling. Bacteria rely on flocculation in order to settle. A critical mass is required to ensure enough contact between the flocculating particles.

Another method to extend the useful life of the system is to divide the aeration basin into two or more tanks. In our example, we could use two 20,000 gallon tanks instead of the one 40,000 gallon tank. Assuming 1,500 mg/l MLSS, at year 6 one aeration basin could be shut down. This would effectively halve the sludge age in the system at a steady MLSS. An added advantage of this method would be that half of the blowers could also be shut down. The system would not only

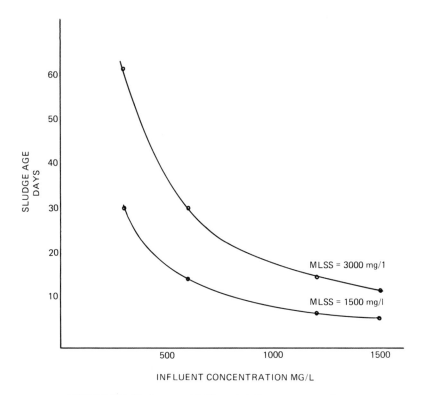

FIGURE 2-3. Sludge age with life cycle influent concentrations.

last longer, but would also cost less to run in the final years of operation.

Of course, there are limitations to an activated sludge system designed this way. The final few years of the cleanup will still create a very long sludge age. The actual design may have to include different unit operations to clean up the groundwater over the entire life-cycle. (See Biological Treatment of a Groundwater Contaminated with Phenol in Chapter 6 for an example of life-cycle design using biological methods). The point is that the change in concentration of contaminants over the life of the project may have a detrimental effect on the performance of the treatment system. The design engineer must take into account the entire range of concentrations when designing the treatment system. Similar limitations will effect fixed film, biological treatment system designs. These problems are not limited to biological systems, inorganic treatment systems also have life-cycle considerations.

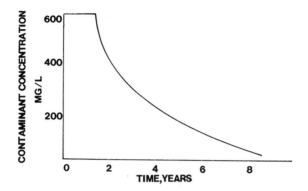

FIGURE 2-4. Life cycle concentration from a well at the center of the plume for an inorganic contaminant.

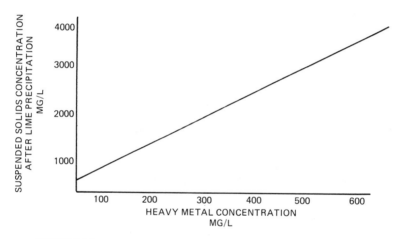

FIGURE 2-5. Suspended solids from lime precipitation of a heavy metal.

For our second example, let us consider an aquifer contaminated with a heavy metal. The design details for removing a heavy metal from water are covered in Chapter 5. For this example, assume that by adding lime, to a pH of 9, all of the heavy metal comes out of solution. Suspended solids are formed, metal hydroxides, that must be separated from the water. Figure 2-4 summarizes the life-cycle concentration for the heavy metal. Assume that laboratory analysis is performed and Figure 2-5 summarizes the concentration of suspended solids resulting from lime precipitation of the heavy metal at various concentrations.

From this data, the design engineer must select the proper unit operation for separation of the suspended solids from the groundwater. During the first couple of years, the concentration of suspended solids will be about 4,000 mg/l. In this range, the main problem with separation will be the thickening of the solids as they are removed from the water. A standard clarifier, with its design based on thickening, is the proper unit operation.

After year three, the standard clarifier is no longer effective for solids removal. Metal hydroxide suspended solids rely on flocculation to be removed from the water. As in activated sludge, a critical mass must be present to ensure enough contacts for proper settling and thickening. As the suspended solids drop below 1,000 mg/l, the preferred separator would be a flocculating clarifier. As the solids decreased to less than 500 mg/l, a solids contact clarifier would be required. The flocculating clarifier has a chamber that increases the number of contacts between floc particles before entering the clarifier zone. The solids contact clarifier allows previously settled solids to contact incoming solids in obtaining a critical mass for settling.

During the final years of the project, the concentration of suspended solids would be so low that a sand filter or a dual media filter would be required to remove the contaminants. The filter would be the proper choice after the suspended solids were below 100 mg/l. Finally, the heavy metal concentration would be at the point that lime precipitation would not be the proper method of treatment. Once the heavy metal concentration is less than 5 mg/l, another technology, such as ion exchange, would be a more cost effective treatment technique. (It should be noted that all of the concentration values given in this example are approximations. Every heavy metal and groundwater will react differently with lime precipitation. Only through laboratory testing can the design engineer determine when these different levels will be reached.)

Separation of the suspended solids is not the only problem that the design engineer must face. Metal hydroxide sludge must be dewatered or solidified before it can be economically disposed of. Once again, the proper unit operation depends on the amount of solids to be processed. Of course, this value will change over the life of the project.

The final treatment plant design must take all the above requirements into consideration. The design engineer must try to maximize the number of years that the treatment plant will function. Figure 2-6 shows one possible design incorporating the life-cycle considerations. Groundwater is brought into contact with lime in a flash mix tank. The liquid/solids mixture is sent to a solids contact clarifier. The water is sent to a filter and the sludge is sent to a thickener.

The solids contact clarifier should be designed to be operated at a high rate. The high solids loading in the beginning years can be quickly transferred to the thickener. The solids that escape because of the high loading will be captured by the filter. During the middle years, the solids contact clarifier will maintain the critical mass necessary for proper settling.

The filter should be of a dual media design. This design can handle a higher concentration of suspended solids. In the beginning years, the filter will remove the solids from the overloaded solids contact clarifier. During the final years, the clarifier can be shut down and the filter will remove all of the suspended solids.

The thickener will handle the high load of solids during the first several years. During the middle and end years, the thickener will act as a storage tank so that the solids dewatering activities can be run on a periodic basis. The solids dewatering will have to use extra manpower and equipment during the beginning years that will then be scheduled during the middle and final years.

Finally, when the concentration reaches a very low level, all the equipment in Figure 2-6 will have to be shut down. An ion exchange

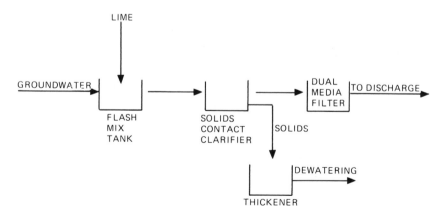

FIGURE 2-6. Life cycle treatment plant design for a heavy metal.

unit will be installed and operated until the last of the heavy metals have been removed from the groundwater.

Of course, Figure 2-6 represents only one possible design. Local conditions may favor a different solution to this design problem. It would be impossible to list all the different designs for this treatment problem. The important concept is for the design engineer to use the life-cycle considerations when designing the treatment plant. These same life-cycle problems will exist on other unit operations. Although, for certain treatment systems, the effect will be more on operational costs than on physical design and possible failure. Carbon adsorption and stripping fit into this category. Both of these unit operations will not fail as the concentration decreases. However, there may be significant cost savings to using other treatment technologies as the concentration goes down. The design engineer must also consider the operational cost over the entire project when developing his design.

Our final example covers the effect on operational costs from a life cycle design. Let us compare the costs of a carbon adsorption system to a packed tower air stripper with vapor phase carbon adsorption. The design criteria for this example will be:

Flow—40 gpm
Benzene—1 mg/l

There are no other contaminants present, and there is nothing that can interfere with the treatment processes, i.e., silt or iron. To keep the example simple, we will use previous published data.[1] If the reader would like more details on the design or costs presented here, they are available from that source.

The carbon system will consist of two 800 pound carbon units in series. We will assume that the capital cost is $4,200. The operating costs for a carbon system mainly come from the cost of replacement carbon, disposal of used carbon, and transportation of the carbon. We will assume that these costs will be $1.25, $1.75, and $1.50 per pound, respectively. The total operating cost will, therefore, be $4.50 per pound carbon. We will assume that the carbon usage rate is 8 pounds of carbon per pound of benzene. The total carbon adsorption costs will be $4,200 for capital and $17.30 per day for operating.

We will assume that the air stripper will be a 14-inch diameter packed tower with 15 feet of packing. The capital cost will be $11,000.

A blower is required for all air strippers, and the operating cost for the blower will be $0.60 per day. The treatment system will also include a vapor phase carbon unit and a heater. The heater is required to maintain the air at below water saturation. This is necessary to obtain higher carbon efficiency in an air stream than is possible in a water stream. The total capital cost of the heater and the vapor phase carbon unit is $7,500. We will assume that the carbon capacity is two pounds of carbon per pound of benzene. Vapor phase carbon costs a little more than liquid phase carbon. Therefore, the total carbon costs will be increased to $5.00 per pound for carbon. Finally, the heater will require 1.9 kw to increase the air temperature to the required levels. The daily costs for the air blower, heater, and carbon will be $0.60, $4.50, and $4.80, respectively. The total cost for the air stripper with vapor phase carbon is $18,500 for capital and $9.90 per day for operating.

Let us now compare these two designs by using constant influent values and by using life cycle design. Figure 2-7 shows the cumulative costs of the two systems assuming constant influent. The carbon system has a lower capital cost but a higher operating cost. While the total costs are lower for carbon in the first few years, the lower operating cost of the air stripper with vapor phase carbon catch up later on. The break even point is 4.8 years. If the remediation was

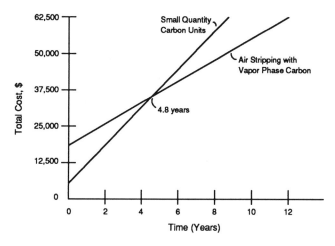

FIGURE 2-7. Cumulative costs of carbon adsorption and air stripping with vapor phase carbon.

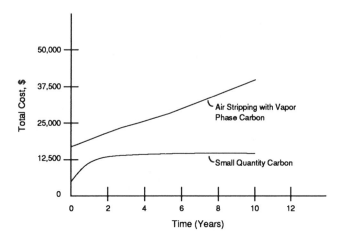

FIGURE 2-8. Life cycle design cumulative costs of carbon adsorption and air stripping with vapor phase carbon.

going to last less than 4.8 years, then the carbon adsorption system would be favored. The air stripper with vapor phase carbon would be favored for longer projects.

The comparison is very different if we use life cycle design. Let us assume that the benzene concentration decreases by 50% per year. At year one the concentration is 0.5 mg/l. At year two, the concentration is 0.25 mg/l, etc. Figure 2-8 summarizes the new cumulative costs. As can be seen, the capital costs remain the same, only the operating costs are affected. The operating costs are lower. However, the decrease in operating costs for the carbon system are so significant, that the lower cost of carbon for the vapor phase carbon system never overcome the capital and other operating costs (i.e. the air heating costs would remain the same, even as the concentration decreases) for that system.

At a minimum, life cycle design must be used to develop accurate costs of treatment for a ground-water remediation. As can be seen in this last example, life cycle design can also affect the type of treatment used for the remediation.

CAPITAL COSTS

In Chapter 1 we saw that the total time for a cleanup would usually be less than the twenty years necessary for a wastewater project. In the previous section of this chapter, we saw that even if the life of the project is 10 years, all of the equipment would probably not be needed for the entire time. In this section, we discuss the effect of time on the cost of equipment.

Most of the equipment used in the field will have a 5 to 20 year life expectancy. Pumps and other equipment with moving parts have a lower life expectancy and tanks and reaction vessels have a longer life expectancy. The cost of equipment in wastewater treatment is figured over the life expectancy of the equipment. The cost of equipment on a groundwater cleanup must be based on the time used on the project with an upper limitation of the life expectancy of the equipment.

Let us assume that the cost of equipment for the activated sludge treatment system example in the previous section was $100,000. If we set the amount of time that we need the equipment and the interest rate that we have to pay for the equipment, then we can calculate the daily cost of the equipment.

One formula for calculating costs would be:

$$C = \frac{\text{Cap}}{[1 - (1 + i)^{-n}]/i}$$

where

C is cost per time period n
Cap is capital cost ($100,000 in our example)
i is the interest rate
n is the period of time

We will assume that the interest rate is 12%. If the equipment is used for ten years, the daily cost is $48/day. If the equipment is only needed for five years, the daily cost is $74/day. At two years, the daily cost is $157/day, and at one year, the daily cost is $296/day. Figure 2-9 summarizes the daily cost of equipment when used for various periods of time.

As can be seen, the cost of equipment gets significantly higher as the time of use decreases. The normal method of comparing the cost of treatment by different technologies is to base the comparison on cost of treatment per gallon of water treated. At a flow of 25,000 gal/day, the cost of treatment goes from $0.00192/gal at ten years to

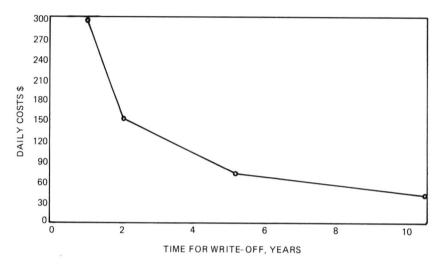

FIGURE 2-9. Capital cost as a function of time.

$0.01184/gal at one year. Using the same equipment for one year will cost six times as much per gallon treated as using the same equipment for ten years.

A great many groundwater cleanups will be completed in a one to two year period. This makes the cost of equipment over time another part of the life-cycle design. The design engineer will have a problem on the shorter projects and on the longer projects in which a particular piece of equipment is only needed for a short period of time. An obvious solution to short term use is to rent the equipment or to use it over several different projects. This would allow the equipment to be capitalized over 10 years even though it was only required for one year on a particular project.

Of course, any equipment that is to be used for more than one project will have to be transported from one site to the next. The equipment will have to be portable. For example, the design engineer has a choice of one tank 17 feet in diameter or two tanks 12 feet in diameter. If the equipment is to be used only a short period of time, the proper choice is the two 12 feet diameter tanks. The legal limit for a wide load on a truck is 12 feet.

In general, to be transported by truck, the treatment equipment should also be less than 10 feet in height and 60 feet in length. Rail transport can take longer units, but to be able to reach most of the U.S., shipment by truck should be assumed in the design.

OPERATOR EXPENSES

One final area that has to be discussed under life-cycle design is operator expenses. Any system that requires operator attention will cost more to operate than a system that does not require operators. All wastewater treatment system designs should have operator expenses factored into their review. With groundwater treatment systems, this factor takes on added importance. The main reason for this importance is the relative size of a groundwater treatment system. Once again, the engineer cannot just take a design developed for wastewater treatment systems and reduce its size for groundwater treatment. Most groundwater treatment systems will be very small in comparison to most wastewater treatment systems. The operator costs, therefore, become more significant.

Let us look at the biological treatment system example once again. Assume that a 15 HP blower is required for the system at $0.06/kwhr. In addition, chemicals and miscellaneous costs are $3.00/day. At a ten year life for the equipment, the daily costs would be:

Equipment	$48.00
Power	$29.00
Chemicals	$ 3.00
Total	$80.00

Figure 2-10 summarizes the relative costs for each category. Without any operator attention, the equipment represents 60% of the daily cost of operation. The power is 36% and the chemicals are 4% of the daily costs. Figure 2-11 shows what happens to this relationship if one operator is required for one, 8-hour shift per day and is paid, including benefits, $10.00/hr. Now 50% of the daily cost is represented by operator costs. Equipment drops down to 30%, power to 18% and chemicals to 2%. At just one shift per day, the operator is now the main expense of the treatment system.

If the treatment system requires full time observation, the operator costs become even more important. Figure 2-12 shows the relative costs when an operator is required 24 hr/day and paid $10.00/hr. Now,

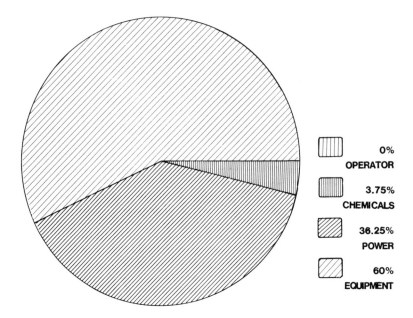

FIGURE 2-10. Ratio of daily costs with no operator attention.

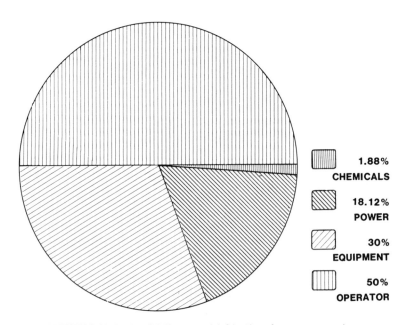

FIGURE 2-11. Ratio of daily costs with 8 hr/day of operator attention.

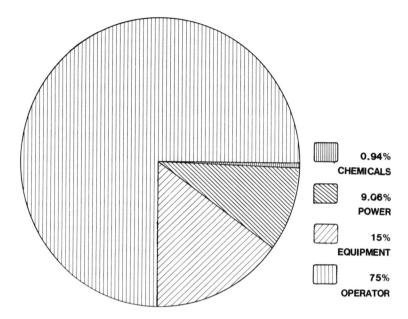

FIGURE 2-12. Ratio of daily costs with 24 hr/day of operator attention.

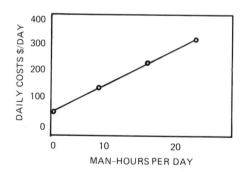

FIGURE 2-13. Daily costs of treatment with variable operator attention.

the operator represents 75% of the cost of operation. Three out of every four dollars spent on the project would go to personnel.

Daily costs for the project double if an operator is required for 8 hr/day when compared to operating with no personnel. The costs triple at two shifts per day, and costs quadruple when around the clock attention is required. These costs are summarized in Figure 2-13. As can be seen from this data, the design engineer can not ignore the effect of the operator on treatment system costs. In fact, the designer should spend most of his effort on minimizing the operator time required for a particular design.

The effect of the operator does not decrease even as the size of the equipment increases significantly. Figure 2-14 represents the relative costs from a treatment system five times the size of the present example and requiring 24 hr/day of operator attention. The operator still represents over one third the cost of treatment. Even as the total cost of the treatment system approaches $500,000 the design engineer must take special precaution to keep the required operator attention to a minimum.

In summary, there are three main factors that must be considered when performing a life cycle design for a groundwater treatment system. First, the concentration may change over time. The treatment design must meet the requirements at the beginning of the project, at the middle of the project and at the end of the project. Second, because of the relatively short time that equipment is needed on

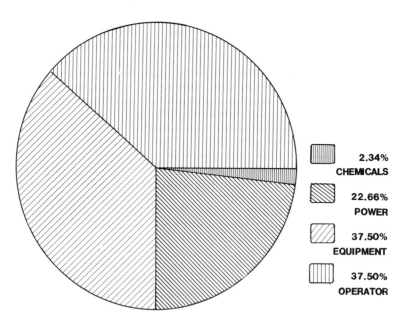

FIGURE 2-14. Ratio of daily costs for a $500,000 treatment system with 24 hr/day of operator attention.

groundwater projects, portable equipment should be considered. Finally, due to the relatively small size of groundwater equipment, manpower costs from operators become a significant, if not controlling factor in equipment design.

Reference
1. Nyer, E. K. The effect of time on treatment economics, *GroundWater Monitoring Review* Spring 1989.

3

Treatment for Organic Contaminants: Physical/Chemical Methods

INTRODUCTION

When a pure organic contaminant is released on to or into the ground, the main force on the movement of the compound is gravity. If the ground is porous, the spill will move downward. There will be some lateral spread of the movement, controlled by the porosity of the soil. The speed of movement will be dependent on the viscosity of the material spilled and the porosity of the soil. Several things can happen to the contaminant as it progresses downward, before it encounters the aquifer. Initially, the contaminant may undergo any of the following: adsorption on the soil particles, volatilization, biological degradation, and usually to a lesser degree, hydrolysis, oxidation, reduction, and dehydrohalogenation.[1] Additionally, the contaminant may encounter an impermeable barrier (clay, bedrock, etc.) which may stop or alter the downward progress.

The adsorption on a soil depends upon both the contaminant and the soil matrix. The solubility, octanol-water partition coefficient (K_{ow}), and the molecular structure of the contaminant are important factors in determining adsorption capacity. In general, the more soluble the compound, the less it will be adsorbed to the soils. The octanol-water partition coefficient represents the distribution of a chemical between octanol and water phases in contact with each other at equilibrium conditions. The higher the octanol-water partition coefficient, the more likely it will adsorb to soils.

The molecular structure relates to the contaminant's[1] polarity and size, both of which directly affect adsorption. These factors are also

44

important if the contaminant reaches the aquifer. Adsorption to soil particles will occur in the aquifer. Adsorption to soil particles will also occur in the aquifer. This is why contaminants move more slowly in an aquifer than in water. The contaminants adsorb and are retarded by the soil particles.

The size of the surface area and organic content are also important soil matrix factors which determine adsorption capacity.[1] The greater surface area associated with clays and clayey materials provides more adsorption capacity. Higher organic content increases the number of potential adsorption sites and therefore increases adsorption capacity.

As the contaminant moves in the soil, each soil particle will adsorb a small amount of the material. The material adsorbed stays with the soil and no longer moves with the main flow of the contaminant. If the spill does not reach an impermeable layer or aquifer first, the spill will eventually exhaust itself in the soil and stop all primary movement.

The amount of soil required to adsorb all of the material released depends upon two factors: the porosity of the soil and the adsorbability of the contaminant as reflected by its characteristic, "maximum residual saturation." When the contaminant is at or below its maximum residual saturation, it will not move in the soil. The American Petroleum Institute[2] recommends the following equation to relate the amount of soil required to immobilize the contaminant:

$$C = (0.20 \times V)/P \times Sr \qquad (3\text{-}1)$$

where:

C = the cubic yards of soil required to immobilize the contaminant
V = volume of contaminant in barrels
P = porosity of the soil
Sr = residual saturation

While this equation was originally developed for petroleum based compounds, the relationships expressed by the formula are valid for other compounds. The material that is adsorbed in the soil has a reduced chance of contaminating the groundwater. The portion of the material that is soluble in water will be picked-up by rain water on its movement through the ground, and acts as a continuing source of contamination. The contaminants that are volatile may also move through the vadose zone and contaminate the aquifer. The rest will remain in the vadose zone.

Without treatment or removal, the adsorbed material can greatly increase the duration of a "pump and treat" recovery program, or if the compound is volatile it can create a direct hazard. Volatile compounds move upward and laterally in the soil entering structures, producing explosion hazards or just nuisance odors.[3] Therefore, the organic material in the vadose must be addressed as part of the remediation. In all pump and treat designs, the source of contamination must be addressed. The adsorbed material in the vadose zone is a potential source of contamination.

The problem lies in the fact that the material adsorbed on the soil particles may be difficult to find and remove. First, there is very little lateral movement in the unsaturated zone. The well must be constructed near the original place where the compound was released, if that site is known. Secondly, in addition to the components of the spill changing due to volatilization and absorption in water, the compounds can also be transformed by (1) the natural biota in the soil, (2) oxidation, (3) reduction, (4) hydrolysis, and (5) dehydro- halogenation. Soil bacteria or these other processes can change the structure of an organic compound. Therefore, the organics found in the soil may be different from the original organics released.

Recently, two methods have been developed for finding material in the vadose zone: soil vapor extraction tests and carbon dioxide measurement.[4] Soil vapor extraction tests detect small amounts of soil vapor over a wide area at a contaminated site. The test will pick up the volatile compounds that exist in the vadose zone. These compounds can be a direct result of adsorbed material or an indication of groundwater contamination. Only volatile compounds are detectable with this method.

A new extension to the soil vapor extraction test is a carbon dioxide measurement. For this measurement, the soil vapor is tested for carbon dioxide instead of organic vapors. It has been shown that the natural biological degradation of the organic compounds results in raised levels of carbon dioxide in the vadose zone. The technical details of why this occurs are covered in Chapter 4. This method is limited to biodegradable compounds, and the results can be masked by natural organic material.

Once found, the material can be removed in several ways. For shallow spills, the ground can be excavated and (1) land farmed (if the organic material is degradable), (2) buried in a controlled landfill, or (3) thermally treated. For deeper contamination, flushing and recovery, vapor extraction systems (VES) or in situ treatment are the main methods of treatment. The biological methods will be covered in the next chapter.

Many of the same problems are encountered when the pure compound movement is stopped by an impermeable layer of soil. When the contaminant reaches the impermeable layer, its downward progress stops. The material spreads out on top of the impermeable layer like a pancake. This process continues until the contaminant contacts enough soil to adsorb the entire spill, or the impermeable layer ends and the contaminant continues downward under the influence of gravity.

The same problems exist for a cleanup of material that hits an impermeable soil layer, as in the cleanup of an adsorbed material. It is hard to gather the material in one spot so that it can be recovered or treated. A slurry wall can be installed to prevent the continued spreading of the contaminant. Also, a well could be drilled down to the impermeable soil and the material recovered. The slurry wall would have to completely surround the spill, which is usually very expensive. The well would not be able to recover a significant portion of the material spilled. Once again, some type of flushing, VES or in situ treatment are the best methods of cleanup.

Finally, the spill can reach an aquifer. Depending on the relative density of the contaminant, the spilled material will either float on top of the aquifer, or sink to the bottom of the aquifer. Additionally, the contaminant solubility will dictate the contaminant concentrations in the ground water.

If the aquifer must be cleaned, the treatment method or methods used for an organic cleanup will depend upon several factors. All of the following will have to be considered when choosing the unit operations to be used:

A. Description of the release
   1. Concentration
   2. Quantity of contaminant
   3. Total time allotted for cleanup, and
   4. Final use of the water.
B. Properties of the spilled material
   1. Solubility
   2. Density

3. Stripability
4. Adsorbability, and
5. Biodegradability.
C. Site and aquifer characteristics
1. Depth to water
2. Permeability
3. Extent of contamination, and
4. Ongoing site activities.

Other factors will be important on individual projects. This chapter addresses the physical and chemical methods of removing organics from an aquifer. Chapter 4 reviews biological methods and in situ treatment. Chapter 5 covers inorganic treatment techniques. Finally, Chapter 6 combines the design considerations presented in Chapters 1 and 2 with the treatment techniques presented in Chapters 3, 4, and 5, to provide examples of groundwater cleanups.

PURE COMPOUND RECOVERY

Pure compound recovery is only possible when the contaminant is not entirely soluble in water. Once the organic is in solution, the pure compound recovery techniques will not work. Very soluble compounds such as acetone and phenol cannot use pure compound recovery techniques. This also includes the situations when a surfactant or similar compound creates an emulsion. Table 3-1 lists the solubility, Table 3-2 lists the specific gravity, and Table 3-3 lists the octanol-water partition coefficient for thirty organic compounds. Henry's law constant (stripability), carbon adsorption capacity, and degradability on these same thirty compounds will be provided in their respective treatment sections.

Straight chain hydrocarbons are generally not soluble in water. Most petroleum products, oil, gasoline, etc., fall into this category. Chlorinated hydrocarbons are also not soluble in water. Chlorinated solvents are the most common insoluble chlorinated hydrocarbons.

If the compound is not soluble, then it will either float on top of the aquifer or sink to the bottom. The compound's relative position to the water in the aquifer will depend upon the relative density of the material. Organic compounds that are lighter than water will float, and organic compounds that are heavier than water will sink.

TABLE 3-1   Solubility for Specific Organic Compounds

	Compound	Solubility (mg/l)	Reference
1	Acetone	$1 \times 10^{6*}$	1
2	Benzene	1.75×10^3	1(A)
3	Bromodichloromethane	4.4×10^3	2
4	Bromoform	3.01×10^3	1(B)
5	Carbon tetrachloride	7.57×10^2	1(A)
6	Chlorobenzene	4.66×10^2	1(A)
7	Chloroform	8.2×10^3	1(A)
8	2-Chlorophenol	2.9×10^4	2
9	p-Dichlorobenzene (1,4)	7.9×10^1	2
10	1,1-Dichloroethane	5.5×10^3	1(A)
11	1,2-Dichloroethane	8.52×10^3	1(A)
12	1,1-Dichloroethylene	2.25×10^3	1(A)
13	cis-1,2-Dichloroethylene	3.5×10^3	1(A)
14	trans-1,2-Dichloroethylene	6.3×10^3	1(A)
15	Ethylbenzene	1.52×10^2	1(A)
16	Hexachlorobenzene	6×10^{-3}	1(A)
17	Methylene chloride	2×10^4	1(B)
18	Methylethylketone	2.68×10^5	1(A)
19	Methyl naphthalene	2.54×10^1	2
20	Methyl tert-butyl-ether	4.8	3
21	Naphthalene	3.2×10^1	2
22	Pentachlorophenol	1.4×10^1	1(B)
23	Phenol	9.3×10^4	1(A)
24	Tetrachloroethylene	1.5×10^2	1(A)
25	Toluene	5.35×10^2	1(A)
26	1,1,1-Trichloroethane	1.5×10^3	1(A)
27	1,1,2-Trichloroethane	4.5×10^3	1(A)
28	Trichloroethylene	1.1×10^3	1(A)
29	Vinyl chloride	2.67×10^3	1(A)
30	o-Xylene	1.75×10^2	1(C)

*Solubility of 1,000,000 mg/l assigned because of reported "infinite solubility" in the literature.

1. *Superfund Public Health Evaluation Manual,* Office of Emergency and Remedial Response Office of Solid Waste and Emergency Response, U.S. Environmental Protection Agency, 1986.
    A. Environmental Criteria and Assessment Office (ECAO), EPA, *Health Effects Assessments for Specific Chemicals,* 1985.
    B. Mabey, W. R., Smith, J. H., Rodoll, R. T., Johnson, H. L., Mill, T., Chou, T. W., Gates, J., Patridge, I. W., Jaber H., and Vanderberg, D., *Aquatic Fate Process Data for Organic Priority Pollutants,* EPA Contract Nos. 68-01-3867 and 68-03-2981 by SRI International, for Monitoring and Data Support Division, Office of Water Regulations and Standards, Washington, D.C., 1982.
    C. Dawson, et al., *Physical/Chemical Properties of Hazardous Waste Constituents,* by Southeast Environmental Research Laboratory for USEPA, 1980.
2. USEPA *Basics of Pump-and-Treat Ground-Water Remediation Technology,* EPA/600/8-901003, Robert S. Kerr Environmental Research Laboratory, March 1990.
3. Manufacturer's data; Texas Petrochemicals Corporation, *Gasoline Grade Methyl tert-butyl ether Shipping Specification and Technical Data,* 1986.

TABLE 3-2 Specific Gravity for Specific Organic Compounds

	Compound	Specific Gravity*		Reference
1	Acetone	.791		1
2	Benzene	.879		1
3	Bromodichloromethane	2.006	(15°C/4°)	1
4	Bromoform	2.903	(15°C)	1
5	Carbon tetrachloride	1.594		1
6	Chlorobenzene	1.106		1
7	Chloroform	1.49	(20°C liquid)	2
8	2-Chlorophenol	1.241	(18.2°C/15°C)	1
9	p-Dichlorobenzene (1,4)	1.458	(21°C)	1
10	1,1-Dichloroethane	1.176		1
11	1,2-Dichloroethane	1.253		1
12	1,1-Dichloroethylene	1.250	(15°C)	1
13	cis-1,2-Dichloroethylene	1.27	(25°C liquid)	2
14	trans-1,2-Dichloroethylene	1.27	(25°C liquid)	2
15	Ethylbenzene	.867		1
16	Hexachlorobenzene	2.044		1
17	Methylene chloride	1.366		1
18	Methylethylketone	.805		1
19	Methyl naphthalene	1.025	(14°C/4°C)	1
20	Methyl tert-butyl-ether	.731		1
21	Naphthalene	1.145		1
22	Pentachlorophenol	1.978	(22°C)	1
23	Phenol	1.071	(25°C/4°C)	1
24	Tetrachloroethylene	1.631	(15°C/4°C)	1
25	Toluene	.866		1
26	1,1,1-Trichloroethane	1.346	(15°C/4°C)	1
27	1,1,2-Trichloroethane	1.441	(25.5°C/4°C)	1
28	Trichloroethylene	1.466	(20°C/20°C)	1
29	Vinyl chloride	.908	(25°C/25°C)	1
30	o-Xylene	.880		1

*Specific gravity measured for the compound at 20°C referred to water at 4°C unless specified otherwise (20°C/4°C).

1. *Lange's Handbook of Chemistry,* 11th edition, by John A. Dean, McGraw-Hill Book Co., New York, 1973.
2. *Hazardous Chemicals Data Book,* 2nd edition, by G. Weiss, Noyes Data Corp., New York, 1986.

Table 3-2 provides the specific gravity of thirty organic compounds. Water is used as the reference compound for the specific gravities. Therefore, any compound with a specific gravity less than 1.00 would be lighter than water, and any compound with a specific gravity greater than 1.00 would be heavier than water. In general, petroleum based compounds are lighter than water, and chlorinated compounds

TABLE 3-3   Octanol Water Coefficients (K_{ow}) for Specific Organic Compounds

	Compound	K_{ow}	Reference
1	Acetone	.6	1(D)
2	Benzene	131.8	1(A)
3	Bromodichloromethane	75.9	2
4	Bromoform	251.2	1(B)
5	Carbon tetrachloride	436.5	1(A)
6	Chlorobenzene	691.8	1(A)
7	Chloroform	93.3	1(A)
8	2-Chlorophenol	145.0	2
9	p-Dichlorobenzene (1,4)	3980.0	2
10	1,1-Dichloroethane	61.7	1(A)
11	1,2-Dichloroethane	30.2	1(A)
12	1,1-Dichloroethylene	69.2	1(A)
13	cis-1,2-Dichloroethylene	5.0	1(A)
14	trans-1,2-Dichloroethylene	3.0	1(A)
15	Ethylbenzene	1412.5	1(A)
16	Hexachlorobenzene	169824.4	1(A)
17	Methylene chloride	19.9	1(B)
18	Methylethylketone	1.8	1(A)
19	Methyl naphthalene	13000.0	2
20	Methyl tert-butyl-ether	NA	
21	Naphthalene	2760.0	2
22	Pentachlorophenol	100000.0	1(B)
23	Phenol	28.8	1(A)
24	Tetrachloroethylene	398.1	1(A)
25	Toluene	134.9	1(A)
26	1,1,1-Trichloroethane	316.2	1(B)
27	1,1,2-Trichloroethane	295.1	1(A)
28	Trichloroethylene	239.9	1(A)
29	Vinyl chloride	24.0	1(A)
30	o-Xylene	891.3	1(C)

NA = Not available.

1. *Superfund Public Health Evaluation Manual,* Office of Emergency and Remedial Response Office of Solid Waste and Emergency Response, U.S. Environmental Protection Agency, 1986.

   A. Environmental Criteria and Assessment Office (ECAO), EPA, *Health Effects Assessments for Specific Chemicals,* 1985.

   B. Mabey, W. R., Smith, J. H., Rodoll, R. T., Johnson, H. L., Mill, T., Chou, T. W., Gates, J., Patridge I. W., Jaber H., and Vanderberg, D., *Aquatic Fate Process Data for Organic Priority Pollutants,* EPA Contract Nos. 68-01-3867 and 68-03-2981 by SRI International, for Monitoring and Data Support Division, Office of Water Regulations and Standards, Washington, D.C., 1982.

   C. Dawson, et al., *Physical/Chemical Properties of Hazardous Waste Constituents,* by Southeast Environmental Research Laboratory for USEPA, 1980.

   D. *Handbook of Environmental Data for Organic Chemicals,* 2nd edition, Van Nostrand Reinhold Co., New York, 1983.

2. USEPA *Basics of Pump-and-Treat Ground-Water Remediation Technology,* EPA/600/ 8-90/003, Robert S. Kerr, Environmental Research Laboratory, March 1990.

are heavier than water. Spills from gasoline stations and oil pipelines will usually be found on top of the aquifer. Of course, the new unleaded gasolines use several types of organic ring compounds (benzene being the basic structure) to enhance the octane rating. These ring compounds are relatively soluble in water, Table 3-1. While the main part of the gasoline will float, the soluble components will enter the water. This is also true for the alcohol part of the "gasohol" gasolines.

Spills of chlorinated hydrocarbons usually come from storage tanks where the compounds are used as solvents or from poor housekeeping practices where solvents were spilled or dumped. Chlorinated organics are soluble in very low concentrations. The more chlorine substitutions, the less soluble the compound, Table 3-1. They will show up in the part per billion range in the aquifer. However, the threshold taste level for these compounds is also very low and people do notice their presence. More importantly, chlorinated hydrocarbons have been shown to cause cancer in laboratory animals. There

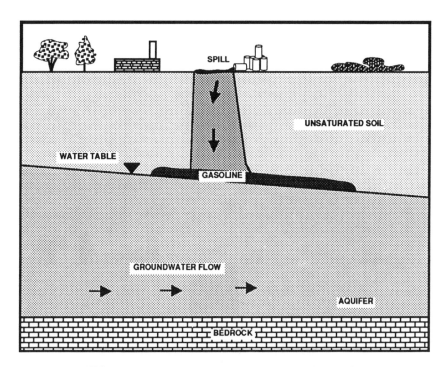

FIGURE 3-1. Petroleum product floating on top of an aquifer.

has been a public outcry about any organics entering the potable water system and current regulations require very low ppb levels. The main part of the compounds will not enter the aquifer, but instead, will continue down through it until they are stopped by an imperme-able soil layer. As with an impermeable layer in the unsaturated zone, the chlorinated compound will then spread like a pancake.

In both cases, the pure compound must be removed, or it will be a continuous source to contaminate the aquifer. The designer should also stress the removal of these contaminants as a pure compound. This is the fastest and least expensive method. By far the easiest compounds to recover are those that float.

The reason that a floating material can be removed from an aquifer with relative ease is that the water in the aquifer can be used to direct the movement of the floating compound. Figure 3-1 shows a floating material on the aquifer. If the water level is lowered in a particular place, the organic compound will continue its original downward path.

Figure 3-2 shows a well being placed in the middle of the spill area.

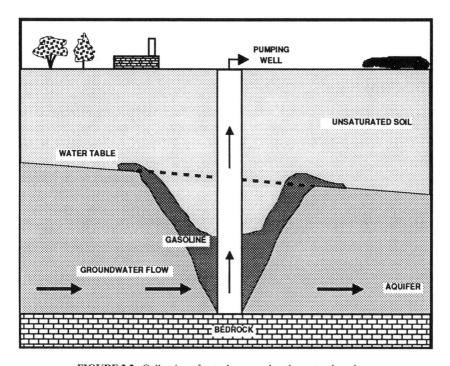

FIGURE 3-2. Collection of petroleum product by water drawdown.

Water is removed, and the water level in the immediate vicinity decreases. The nonsoluble compound will follow the lowered groundwater, and concentrate at the lowest point. The lowest point will be the well. The well can then also be used to remove the floating material. Water should be removed at a fast enough rate to control the entire spread of the floating material.

Once the floating material is accumulating in the well, there are several methods to remove it. The first is to lower a bailer into the well and to let the floating material pour over the top. The lip of the bailer is kept above the water level so that only the contaminant is removed. This process is time consuming and is manpower intensive. Several manufacturers have automated this process.

The bailers have been weighted so that they float in water and sink in anything lighter than water. The same unit can contain a pump for the water to maintain a depression and a timer to raise and lower the bailer. Once the well is drilled and the equipment is set up, only periodic visits to check the equipment are required.

Figures 3-3 and 3-4 show an even more advanced method of removal. This system, once again, uses a pump to maintain the water level. An oil/water separator is lowered into the well. A screen in the separator is coated with a hydrophobic material. The screen will not allow water to pass, but a petroleum based product like gasoline can enter the separator. The screen surrounds a second pump that removes the pure compound from the well. This system can remove greater quantities of material.

Another method combines concepts from the first two methods and can be used with or without a water table depression pump. A pneumatically driven skimmer device has been designed to sink in anything lighter than water and to float just above the water table. The intake which floats just above the water table is made of a hydrophobic material to prevent water from entering it. A pneumatic bladder pump is set up to operate from essentially 0 gpm to 5 gpm. The float is suspended from a coil of tubing to allow the float to rise and fall with variable water table conditions.

When the material is too viscous to move through the screen or when there is a large amount of pure contaminant, a third method can be used. This method requires a large diameter well. Once again two pumps are used. In this case, the pure compound pump is controlled by a conductivity probe. Water carries an electrical charge between

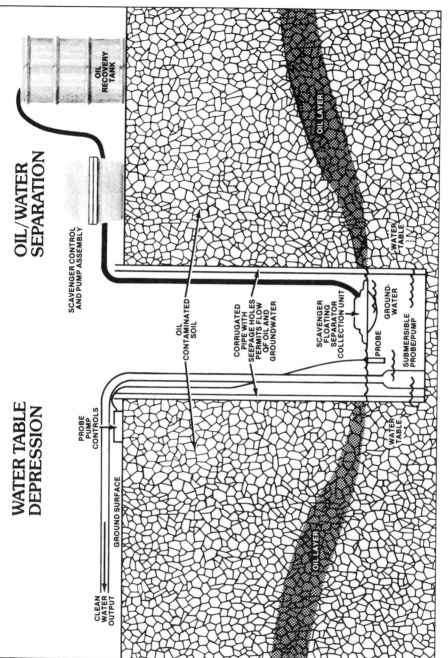

WATER TABLE DEPRESSION

CLEAN WATER OUTPUT

GROUND SURFACE

PROBE PUMP CONTROLS

OIL LAYER

WATER TABLE

OIL/WATER SEPARATION

SCAVENGER CONTROL AND PUMP ASSEMBLY

OIL RECOVERY TANK

OIL CONTAMINATED SOIL

CORRUGATED PIPE WITH SEEPAGE HOLES PERMITS FLOW OF OIL AND GROUNDWATER

SCAVENGER FLOATING SEPARATOR COLLECTION UNIT

PROBE

GROUND-WATER

SUBMERSIBLE PROBE/PUMP

OIL LAYER

WATER TABLE

FIGURE 3-3. Oil/water separation in the well. (Courtesy of Oil Recovery Systems, Inc.)

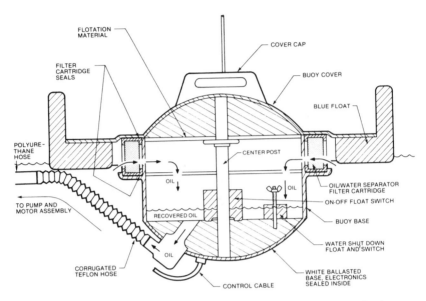

FIGURE 3-4. Oil/water separator. (Courtesy of Oil Recovery Systems, Inc.)

two electrodes. Pure organic compounds do not conduct electricity. When the probe is immersed in a pure compound, no charge passes between the two electrodes of the probe. The unit then knows that it is in a pure compound and turns on the second pump. Figure 3-5 shows a conductivity removal method. One problem with this method is that variable water table conditions can require that pure product pumps be raised or lowered throughout their operation.

One final common method is used for recovery of floating product. For wells with small diameter less than or equal to 4 inches, that yield low flows (i.e., less than 5 gpm), a pneumatic submersible pump can be used. These pumps can be set three ways: (1) to recover floating pure compounds (intakes located at the top of the pump); (2) to recover "total fluids" (intake located at both the top and bottom of the pump), which is a combination of both pure compounds and ground water; or (3) to recover sinking compounds (intake located at the bottom of the pump). If the pump is designed to recover total fluids, a phase separator device (i.e., oil/water separator) must be incorporated into the treatment system.

All five methods have been used on many installations. Several manufacturers have a broad base of experience with each of these

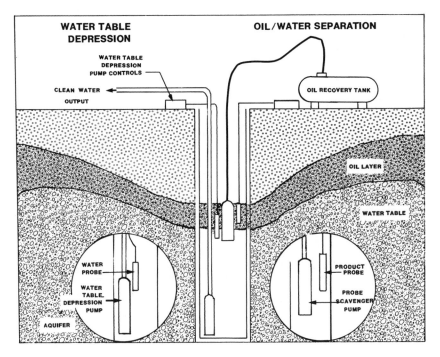

FIGURE 3-5. Oil recovery with a product pump controlled by a product probe.

methods. One of these companies should be contacted to assist on pure compound recovery.

One other way to remove nonflammable pure compounds that float is to dig a trench in the path of the groundwater movement, Figure 3-6. This technique works best for high water tables and other conditions (i.e. perched water tables) where the spill stays near the surface. As the groundwater enters the trench, the floating organic material remains on top of the water. However, once in the trench, the floating material can be removed by skimmers, or other standard, surface removal devices. The water level in the ditch can be lowered to increase the rate of movement of the floating material. Once again, the water in the aquifer is used to control the movement of the pure compound. Special designs of this method must be considered if the pure compound is flammable or if it represents an explosion hazard.

Pure compounds that sink cannot be recovered by using the aquifer to control their movement. This makes their recovery very difficult. The first problem is to find the pool of pure material. A test well must

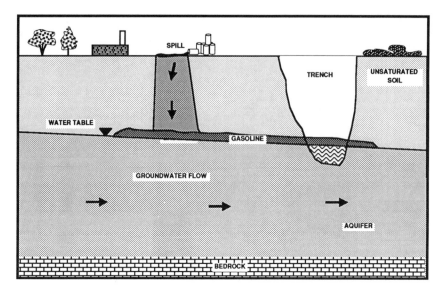

FIGURE 3-6. Removal of floating material by an intercepting trench.

be drilled directly into the material. Once the material is found, the only way to control the removal is to use the conductivity method or "total fluid recovery" using a pneumatic submersible pump, and then separation in an above ground separator. With the conductivity method, only one pump is used, and it is controlled by the nonconductance of the pure organic compound.

Chlorinated hydrocarbons, and other heavier than water compounds, may or may not move in the direction of the groundwater. Therefore, a change in the groundwater flow may or may not affect the movement of the chlorinated compound. In particular cases when wells can be used to force the chlorinated hydrocarbons in a particular direction, the process will not be as efficient as with floating material. In some cases, the compound can travel in the opposite direction to that of the groundwater flow. Figure 3-7 shows a case where the impervious layer at the bottom of the aquifer slopes in the opposite direction to that of the groundwater flow. The chlorinated hydrocarbon will continue to be affected by gravity and to travel with the contour of the aquifer bottom and not necessarily in the direction of the groundwater flow.

In summary, when a pure compound is released to the ground, we

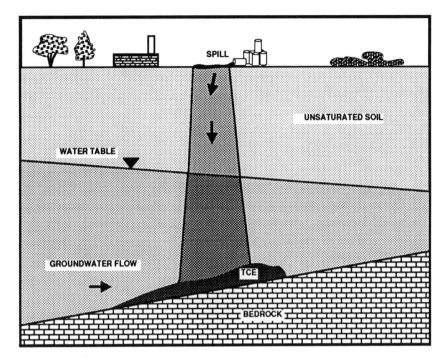

FIGURE 3-7. Chlorinated hydrocarbon movement in opposite direction of groundwater flow.

must consider several important factors in order to cleanup the release. We must first determine if the material has reached the aquifer. We then must discover if the material has stayed on top of the aquifer, dissolved into the water of the aquifer, or passed down through the aquifer. If the material is not in the aquifer, then one of the pure compound treatment methods should be used to recover as much of the material as possible. The remainder of this chapter and Chapter 4 cover treatment technologies for managing the material entering the aquifer.

AIR STRIPPING

Aeration has been used for decades for the removal of dissolved gases such as carbon dioxide, hydrogen sulfide, and ammonia from water and wastewater. It has also been used to introduce oxygen as a means of increasing dissolved oxygen content and for the oxidation of dissolved metals such as iron. It was not until the mid 1970s that the technology was applied for the treatment of water with low levels of volatile (synthetic) organic contaminants.[5]

Aeration relies on the exposure of the contaminated water to a fresh air supply. As the air and water mix, the volatile compounds in the water are driven out of solution and into the vapor state. Maximizing air/water contact is the key to any aeration system. There are many ways to achieve this goal. Numerous technologies are listed in Table 3-4. Among these, packed tower aeration (PTA) is the most commonly applied for the removal of VOCs from water.[6] The acceptance of PTA is reflected in the United States Environmental Protection Agency's (USEPA) acknowledgement of air stripping as a Best Available Technology (BAT) for the treatment of VOCs.[7] The USEPA selection of PTA as a BAT is based on the degree of treatment that can be achieved with this technology when it is used for waters contaminated with the most commonly occurring VOCs.

While the government and industry have settled on PTA as the main type of air stripper employed, it is important to remember that

TABLE 3-4   Common Aeration Technologies

Device References	Typical Configuration	Removal*	Cost($/1000 gal)
Slat tray aerator	Redwood slat trays in box-like structure 10 to 18 feet tall. Countercurrent air/water flow.	60% to 80%	$0.05
Diffused air	Water storage basin with air diffusers. Up to 20 minute contact times common.	< 90%	$0.40 to $2.00
Spray aeration	Spray nozzles in open or closed system. Fine droplets provide air/water contact.	50% to 90%	NA
Cascade aerator	Exposed system of stacked trays, relies on natural draft.	50%	$0.05
Packed column	Cylindrical with plastic media. Countercurrent air/water flows.	90% to 99.9%	$0.05 to $0.25
Rotary stripper	Rotating packed bed. Relies on centrifugal force to form thin liquid films and high turbulence.	> 90%	NA

*Removal of Trichloroethylene.
NA = Not available.

PTAs are only one type of air stripper. Specific situations may require other aeration designs as the best technical and economic choice of treatment system. Small flows (<25 gpm) and high iron content are two examples of factors that may lead to alternative designs. However, since PTA is the method of air stripping that has found the most acceptance for both potable water purification and remedial work on groundwater contamination, the majority of the discussion that follows will center on PTA. Much of what follows is also equally valid for other methods of air stripping. We will discuss and compare the other types of air strippers at the end of this section.

Design of Packed Towers

Air stripping is a mass transfer process. In a packed tower aeration system, air and water are run counter-current through a randomly-dumped or structured media. The media enhances air/liquid contact by breaking the water into thin films and exposing a large amount of liquid surface area to the air. The more surface area exposed, the greater the opportunity for transfer of the volatile organics out of the water into the passing air. The media also serves to continually mix the water so that the stripping process is not limited by diffusion of the VOCs through the water. The air carries the contaminants out of the stripper and into the atmosphere where the volatile organics are dissipated by the velocity of the air stream and by any wind currents, and where the chemicals may breakdown by natural UV degradation or other methods (chemical, biochemical, etc.). The treated water passes out of the column for use, discharge, or further treatment, if required, as depicted in Figure 3-8. In some situations, local regulations and/or the level of VOCs emitted in the contaminated air stream necessitate the use of treatment technologies for the air stream.

When sizing a packed column, the design engineer has three basic variables to define: (1) tower height; (2) tower cross-section; and (3) air to water ratio. While these variables are dependent upon each other (i.e., a change in air to water ratio may allow, or require, a change in packing height), the following basic relationships are helpful in preliminary sizing estimates:

Tower cross-sectional area is most strongly a function of water flow rate. The cross-sectional area of a tower will be determined by the

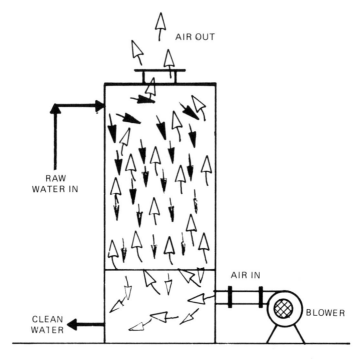

FIGURE 3-8. Packed tower air stripper.

flow rate and the liquid loading rate, the latter commonly ranging
between 15 and 35 gpm/sq. ft.
Tower height is most strongly a function of the removal efficiency
required; the greater the efficiency required, the taller the tower.
The treatment efficiency for a given incremental bed depth is
constant, i.e., if 10 feet of packing achieves 90% removal, an
additional 10 feet will achieve 90% of what remains, for a total of
99% removal.
Air to water ratio is a function of the contaminant being removed.
The more volatile a substance is, the smaller the volume of air
required to strip that compound. Air to water ratios range from
10:1 to 200:1 depending on the compound being removed.

Another important variable in air stripping design is the water
temperature. Water temperature influences the air stripping process
in that higher temperatures enhance constituent volatility. In general,

the design engineer must be aware of the temperature of the water that is to be treated. Groundwater temperatures vary throughout the country and pretreatment equipment may raise or lower the temperature of the water entering the air stripper. Ambient temperature strippers experience little, if any, change in operating performance between winter and summer operation, although air temperatures may fluctuate over a 100°F range. This is because the groundwater temperature stays at or near a constant temperature year round, and the thermal mass of the water is much greater than the thermal mass of the air in the stripper. Thus, the actual operating temperature of the stripper remains fairly constant.

The temperature may be changed through the use of preheaters on the water stream or by the injection of steam directly into the tower. Stripping will occur at a higher rate at elevated temperatures, and some compounds that are barely volatile at ambient temperatures can be totally removed by raising the column temperature into the 140-180°F range.[8] The use of high temperature strippers or steam stripping is generally limited to hazardous waste site cleanups of short duration, or in situations with high levels of organics where air treatment is required. The limitation of high temperature or steam strippers is chiefly due to the high operating costs associated with heating the process streams.

The design method of any packed column starts with the basic mass transfer process. The rate of transfer of the volatile organic compound will be a function of the driving force (the concentration gradient between water and air) and the air/water interface area. Different compounds will be transferred at different rates, depending on the contaminant's Henry's law constant. The Henry's law constant is the ratio of the partial pressure of a compound in air to the mole fraction of the compound in water at equilibrium. Compounds with a high Henry's law constant have a greater concentration in air when an air/water system is in equilibrium. These compounds undergo a phase change from the dissolved state to vapor quite easily, and hence are easily stripped. Compounds with low Henry's law constants, on the other hand, are more hydrophilic and are more difficult to strip. Table 3-5 provides the Henry's law constants for thirty organic compounds at a fixed temperature. The Henry's law constant varies with temperature. Table 3-6 lists the functional relationship of the Henry's law constant and temperature for several common VOCs.

TABLE 3-5   Henry's Law Constants for Specific Organic Compounds

	Compound	Henry's Law Constant[a] atm	Reference
1	Acetone	0	1
2	Benzene	230	1
3	Bromodichloromethane	127	1
4	Bromoform	35	3
5	Carbon tetrachloride	1282	1
6	Chlorobenzene	145	2
7	Chloroform	171	1
8	2-Chlorophenol	0.93	2
9	p-Dichlorobenzene (1,4)	104	4
10	1,1-Dichloroethane	240	1
11	1,2-Dichloroethane	51	1
12	1,1-Dichloroethylene	1841	1
13	cis-1,2-Dichloroethylene	160	1
14	trans-1,2-Dichloroethylene	429	1
15	Ethylbenzene	359	1
16	Hexachlorobenzene	37.8	2
17	Methylene chloride	89	1
18	Methylethylketone	1.16	2
19	Methyl naphthalene	3.2	2
20	Methyl tert-butyl-ether	196	1
21	Naphthalene	20	4
22	Pentachlorophenol	0.15	2
23	Phenol	0.017	2
24	Tetrachloroethylene	1035	1
25	Toluene	217	1
26	1,1,1-Trichloroethane	390	1
27	1,1,2-Trichloroethane	41	2
28	Trichloroethylene	544	1
29	Vinyl chloride	355000	3
30	o-Xylene	266	1

[a] = at water temperature 68°F

1. Per Hydro Group, Inc., 1990.
2. Solubility and vapor phase pressure data from *Handbook of Environmental Data on Organic Chemicals,* 2nd edition, by Karel Verschueren, Van Nostrand Reinhold, New York, 1983.
3. Michael C. Kavanaugh and R. Rhodes Trussel, "Design of aeration towers to strip volatile contaminants from drinking water", *Journal AWWA,* December 1980, p. 685.
4. Coskun Yurteri, David F. Ryan, John J. Callow, Mirat D. Gurol, "The effect of chemical composition of water on Henry's law constant", *Journal WPCF,* Volume 59, Number 11, p. 954, November 1987.

TABLE 3-6   Selected Henry's Law Constants As a
Function of Temperature

$H(atm) = 10^{\wedge(b-a/T)}$ (T in Kelvin)

Compound	a	b
Trichloroethylene	1716	8.59
Tetrachloroethylene	2159	10.38
Benzene	1852	8.68
Toluene	1492	7.427

The mass transfer equations for an air/water stripping system are:

$$Z = HTU \times NTU \tag{3-2}$$

$$HTU = L'/K_{la} \tag{3-3}$$

$$NTU = (R/(R-1)) \times \ln\left[((C_{inf}/C_{eff}) \times (R-1) + 1)/R\right] \tag{3-4}$$

where:

$$R \quad = (H \times G)/(L' \times P)$$

and

HTU = Height of transfer unit (feet)
NTU = Number of transfer units (unitless)
H = Henry's law constant (atm)
G = Gas loading rate (cfm)
L' = Liquid loading rate (gpm/sq. ft.)
P = Operating pressure (atm)
Z = Packing height (feet)
C_{inf} = Influent concentration ($\mu g/l$)
C_{eff} = Effluent concentration ($\mu g/l$)
K_{la} = Overall mass transfer coefficient (sec^{-1})
R = Stripping factor (unitless)

(Note: Unit conversion factors are not shown in the above equations.
Units should be consistent.)

The key variables to define in the equations above are Henry's law constant (H) and the overall mass transfer coefficient (K_{la}). The Henry's law constant is available from a variety of sources; however, care must be taken in using the Henry's law constants, since published values of Henry's law constants can vary by more than an order of magnitude. Additional theoretical research needs to be done in this area.

The mass transfer coefficient is a function of tower design and type of packing, and is a good expression of the overall efficiency of the tower. Accurate quantification of this coefficient is very important, since it can be seen from Equations 3-2 and 3-3 that there is an inverse relationship between tower height and K_{la}. Thus, a 25% error in a K_{la} value might add 25% to the height of the tower, resulting in increased costs due to additional shell material and packing. Because of this relationship, it is good engineering practice to select a mass transfer coefficient based on some type of field data, such as pilot test results or operating data from a similar installation on a similar water supply. Figure 3-9 shows a typical pilot system configuration. A pilot study on the actual water to be treated is best, since chemical characteristics (such as the type and amount of contamination) will vary from source to source and may affect the stripping process. Pilot testing can be simple and inexpensive; usually enough data for design purposes can be obtained in a single day of testing. The testing should cover a range of possible liquid loading rates and air to water ratios, as the K_{la} will vary with both of these factors. The engineer selects a Henry's law constant value for the compound to be tested. He then varies the gas and liquid loading rates and measures the resulting effluent concentration for a constant influent concentration. The mass transfer equations can then be solved for the K_{la}.

When field data is not available or pilot testing is impractical or expensive, the use of theoretical correlations is acceptable. Correlations such as those developed by Onda, Sherwood and Holloway and Shulman[9] are commonly applied. These correlations were developed utilizing bench scale data to produce best fit curves to describe the relationship between the K_{la} and physical properties of the air, water, organic compounds, and packing.

Published reports from a variety of sources [9,10,11,12,13] suggest that the

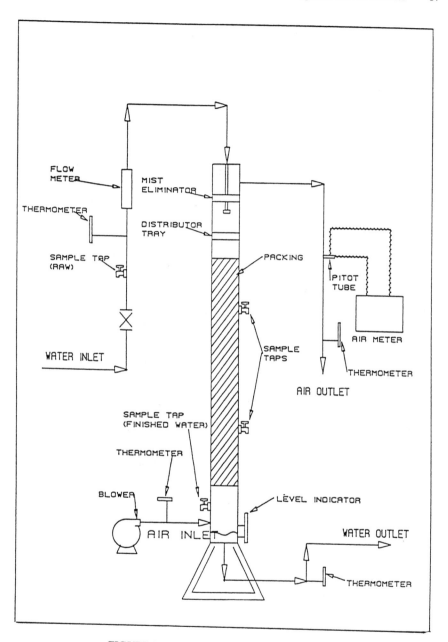

FIGURE 3-9. Pilot plant packed tower air stripper.

Onda correlation is an acceptable means of evaluating air stripper designs for an air/water countercurrent system. In one study, the Onda correlation provided the most accurate correlation of the three theoretical correlations mentioned above.[9] In a comparative study utilizing published pilot scale data, the Onda correlation was found to provide a mass transfer rate (K_{la}) with an accuracy of 30%.[13] In a later study,[5] the Onda correlation overestimated the K_{la} in 75% of the situations by an average of 37%. (For those cases, the Onda correlation would underestimate the packed bed depth and the packed tower would fail to perform as designed). The variations between Onda and the empirical data supports the argument that adequate safety factors should be applied to any designs generated using the correlation.

When the Henry's law constant is known and a K_{la} determined for the range of water loading rates and air to water ratios of interest, then Equations 3-2, 3-3, and 3-4 may be solved for different combinations of packing height and liquid loading rates. The design engineer can use this data to generate various tower configurations that will provide the required removal efficiencies. Some of these options can be eliminated due to site-specific restraints, such as a maximum allowable height. After these are removed, an estimate of capital costs and operating costs should be made for each tower, and a final tower design selected.

Cost of treatment can vary widely for packed towers, depending on the removal efficiency required, the compounds involved, the supporting equipment required, and the need for air treatment. Without air treatment the costs range from approximately $ 0.04/1000 gallons for a large unit treating a fairly volatile organic material to $ 0.17/1000 gallons for a stripper that requires a very high air to water ratio, and hence has high operating costs due to the energy requirements of the blower. If off-gas treatment is necessary, costs will increase by as much as one to two dollars per 1000 gallons treated.

Column Components

The major components of a stripping tower are the tower shell, tower internals, packing, and air delivery systems as shown in Figure 3-10. The tower shell is usually cylindrical, for strength, ease of fabrication, and to prevent channeling of the air or water. Square or rectangular towers, while not common, are used particularly in situations where

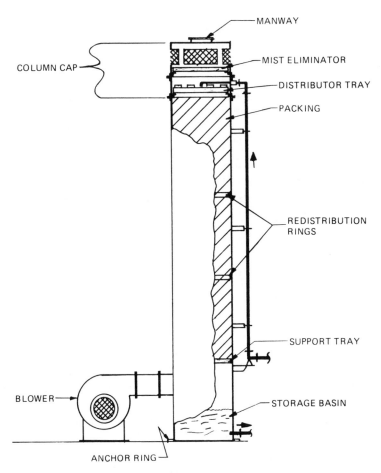

FIGURE 3-10. Packed tower components.

the tower must be housed or when the tower shell is concrete. The tower must be built to withstand all applicable wind, snow, and earthquake loads for the area in which it is being installed, and must be able to support the combined weight of the tower internals, packing, and the water held up in the tower. However, the tower does not need to be designed to support a full column of water, since it will never be entirely filled with water during operation. Air openings at the bottom of the tower will allow water to drain out in case of an obstruction in the effluent water line, thereby acting as a safety outlet to prevent column flooding.

Materials of shell construction include aluminum, fiberglass, stainless steel, coated carbon steel, and concrete. Fiberglass and aluminum are the least expensive materials, with stainless steel and coated carbon steel being slightly higher in cost in most cases. Concrete towers, typically with a brick face, are the most expensive. It is important to remember that raw material prices can fluctuate rapidly, so all applicable materials should be considered at the time of tower design. Costs aside, the relative advantages of the various materials hinge on their strength and corrosion resistance. Aluminum has excellent structural properties, is lightweight, and is suitable for potable water applications. It should not be used on highly acidic (pH <4.5) process streams, or where large amounts of chlorides or heavy metals are present. Fiberglass towers offer good corrosion resistance in most chemical environments (the hand lay-up method is preferred to filament wound columns for chemical resistance), but is comparatively brittle, making it difficult to construct tall towers which are subject to wind loads.

The various stainless steel alloys offer a wide range of corrosion resistance, as well as good structural properties, but many alloys are not readily available without premium prices. Carbon steel, with an epoxy coating offers corrosion resistance and strength at low cost, but requires increased maintenance costs for painting and periodic internal inspection.

Towers constructed of concrete are field erected. They provide high quality aesthetics at a premium price. These units are also subject to honeycombing and leaks if the concrete is not properly poured. Table 3-7 summarizes the properties of the various materials of construction.

The tower internals serve to ensure that the mass transfer process takes place under optimal conditions, at the most economical cost. Starting at the top of the tower, the first component that requires a design engineer's attention is the air exhaust ports. (For the purposes of this discussion, a forced draft tower will be discussed. Induced draft towers will be explored later.) These ports are typically located around the circumference of the tower, and sized to permit the air to escape with a minimum pressure drop. If the tower is for potable water, the outlets should be screened to prevent contamination by wind-borne material entering the tower; towers screened with 24

TABLE 3-7   Packed Column Air Stripper—Materials of Construction

Material	Advantages	Disadvantages
Aluminum	—Lightweight —Low cost —Corrosion resistant —Excellent structural properties —Long life (>15 years) —No special coating required	—Poor resistance to water with pH less than 4.5 and greater than 8.6 —Pitting corrosion will occur in the presence of heavy metals. —Not well suited to high chloride water.
Carbon Steel	—Mid-range capital cost —Good structural properties —Long life if properly painted and maintained	—Requires coating inside and outside to prevent corrosion, leading to increased maintenance. —Heavier than aluminum or FRP.
Fiberglass	—Low cost —High chemical resistance to acidic and basic conditions, chlorides and metals	—Poorly defined structural properties. —Short life (<10 years) unless more expensive resins used. —Poor resistance to UV light (can be overcome with special coatings that must be maintained. —Requires guy wires in most situations. —Susceptible to extremes of temperature differential disturbing tower shape and interfering with distribution.
Stainless Steel	—Highly corrosive resistant Excellent structural properties Long life (>20 years) No special coating required	—Most expensive material for prefabricated towers. —Susceptible to stress fracture corrosion in the presence of high chloride levels.
Concrete	—Aesthetics —Less prone to vandalism	—Difficult to cast in one place leading to potential difficulties with cracks and leaks. —More expensive than self-supporting prefab towers.
Metal lined block and brick	—Aesthetics —Less prone to vandalism —Prefab air stripper insert eliminates problems associated with cast in place towers	—More expensive than self-supporting prefab towers.

mesh screen have reported no problems in this regard. Towers treating large quantities of heavily contaminated water may require tall stacks to direct the exhaust air up and away from the immediate area or more duct work to channel the offgas to an air treatment device. This is more fully discussed in the section on air treatment.

Continuing downward, the next component encountered is the mist eliminator system, placed in the tower to prevent the discharge of large quantities of water entrained in the air stream. This is accomplished by forcing the air stream through a series of bends to impinge the water droplets on the surface of the mist eliminator. Mist eliminators come in two broad categories: chevron-type and pads. The chevron-type eliminators are made of angled plates placed next to each other such that the air is forced to "zigzag" through, impinging the water on the plates. The chevron baffles are typically constructed of PVC, fiberglass reinforced plastic (FRP), stainless steel, or aluminum. The baffles typically provide 4 to 12 direction changes as the air flows through. The deeper the baffle arrangement and the tighter the centerline distance between baffles, the better the mist elimination efficiency.

The pad type mist eliminators are composed of filaments loosely bundled or woven into pads ranging from 3-inches to 12-inches thick. As the air flows through the pad, the water droplets are deposited on the filaments, which are usually made of polypropylene or stainless steel. The pad type mist eliminators can provide greater mist removal, especially at higher air flow rates. However, they have a higher pressure drop. In addition, the pad type mist eliminators tend to be less expensive.

Water is introduced into the tower by means of a water distributor, which ensures that the water is evenly distributed across the surface of the packing and the cross-section of the tower, while allowing for smooth, unimpeded airflow upward to the top of the tower. The distributors fall into four general categories: orifice pans, Figure 3-11; weir and trough arrangements, Figure 3-12; header-lateral piping; and spray nozzles. The header-lateral and weir and trough systems rely on the same basic concept, dividing the flow into successively smaller streams. The major drawback of these systems is the difficulty in assuring even water distribution, a factor that is critical to efficient tower operation. Weir systems have certain "blind" spots under the troughs where water does not fall, and header lateral systems are

FIGURE 3-11. Orifice pan distributor.

FIGURE 3-12. Weir and trough distributor.

notorious for unequal flow in different laterals, depending on their location. However, these systems do find use where large air flows are required, since they provide a large open area through which the air can pass. The weir and trough distributor has the unique advantage of providing for large turn down ratios (the ratio of the maximum flow and the minimum flow that the tower will see). These distributors can be designed for turn downs as high as 7:1. Header lateral units are typically not used in situations where variable flow (turn down ratios greater than 1:1) is expected. The turn down ratio becomes important for variable flow situations. When the groundwater is affected by rain and other outside influences, the flow from the wells may have to vary in order to maintain an adequate capture zone. The design of the treatment system, including all of its components, will have to include the possible variation in flow.

Orifice-type distributors trays avoid these problems of unequal distribution. The trays are designed to keep a standing head of water in them, thereby assuring that an equal pressure, and hence an equal flow, will be maintained at each orifice. Air stacks are provided to allow the gas flow through the tray. These trays do an effective job, and are generally less expensive to fabricate than other distributor systems. The trays are typically designed to provide for turndown ratios as high as 3:1.

Spray nozzles are used most often in low-cost, off the shelf units. The major advantage of spray nozzles is that they immediately break up the water flow into droplets, thereby enhancing the mass transfer process. The major drawbacks are: increased water pressure required to operate them, typically 2 to 5 psi, resulting in increased pumping costs; the extra tower space required to allow for their use; and clogging of the nozzle, especially in turbid waters. While spray nozzles can be designed for turndown ratios as high as 2:1, the operating cost associated with this capability is typically an additional 5 psi pressure loss.

Several other components are involved in maintaining even air and water distribution in the tower. Throughout the depth of the packing are wall wipers and/or redistributors, which serve to rechannel any wall flow back toward the center of the column and ensure consistent distribution of air and water throughout the packed bed. Wall wiper rings should be provided at approximately 5 foot intervals. Redistributors, usually orifice pan or orifice plate distributors should be provided at least every ten diameters of packed bed depth and sometimes as often as every six diameters. Thus, if a 25-foot bed is required in a 2-foot diameter tower, at least one redistributor and possibly two should be provided.

Below the distributor lies the packing, which is held up by a packing support plate. We will finish discussing the structural components before moving on to the packing. The support plate must be structurally capable of supporting not only the weight of the packing, but also the weight of any water present in the packed bed and any inorganic buildup that may collect on the packing over the life of the stripping unit. At the same time, the plate must have enough open area to deter flooding, a condition that results when the water flow downward through the tower is significantly impeded by the upward gas flow. When the water flow is restricted, a head of water builds up until the water's weight is enough to force its way through the plate. This leads to an unstable "burping" action, where first water and then air are alternately passed through the plate, decreasing removal efficiencies in the tower.

For most water treatment applications, an FRP, aluminum stainless steel, or polypropylene grating will provide adequate open area to prevent flooding, approximately 70% open area. In designs with very high liquid and gas loading rates, a gas injection plate is sometimes employed. These plates have a wavy appearance, which provides more open area than is possible with a flat plate by as much as 100% of the tower cross-sectional area. Air is "injected" through the sides of the undulations of the plate. These plates are usually fabricated out of stainless steel, and are more expensive than FRP grating.

Whether gratings or air injection plates are employed, the support tray must be held circumferentially and in many cases laterally. This support takes the form of a circumferential ring and structural channels or beams. Designing for a minimum ¼-inch plate deflection will usually insure adequate strength and rigidity.

The design of the tower base will vary with system configuration; an integral clearwell may be supplied as part of the tower, or the water may flow by gravity to discharge in a stream or sewer. Whatever the configuration, it is imperative that a water seal be provided in the discharge line, to prevent short-circuiting of the tower by the air blower out the discharge line. A hydraulic analysis of the discharge should also be performed to ensure that the water will not backup in the tower, possibly flooding the air blowers. Several common configurations are shown in Figure 3-13.

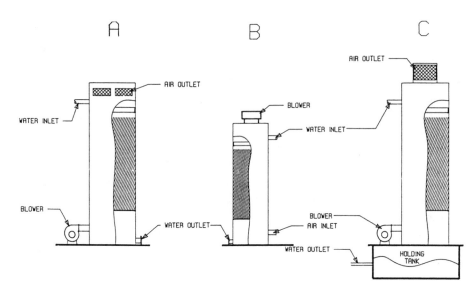

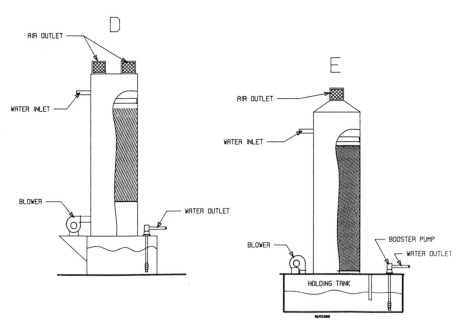

FIGURE 3-13. Common tower configurations. Clockwise: (A.) Integral basin with gravity discharge. (B.) Induced draft system with integral basin and gravity discharge. (C.) Below grade storage basin and booster pump discharge. (D.) Above grade integral booster pump basin. (E.) Low profile below grade basin with booster pump discharge.

The single most important component selection for the design engineer is the tower packing, Figure 3-14. The ideal tower packing will provide a large surface area for the air and water to interact, and will also create turbulence in the water stream, to constantly expose fresh water surfaces to the air. The packing should have a large void area to minimize the pressure drop through the tower. Additional considerations on packing choice include weight, corrosion resistance, the ability to maintain a uniform liquid flow, and of course, price. Table 3-8 summarizes the basic properties of several representative packings. Because new packings are developed constantly, the reader can only use Table 3-8 as a guide, and maintain a literature file on new packings.

Unfortunately, there is no single measure to determine the best tower packing. Measurements such as surface area per unit volume can be misleading because the surface area of the packing is not the same as the air/water interface area. Many of the packings with wide, sweeping surfaces may only have one side of their area wetted, essentially wasting the dry half. Other packings, because of their configuration, may cause channelling of the water, reducing the air/water contact area. For this reason, a comparison pilot-scale test using various packings is the only valid method of evaluation.

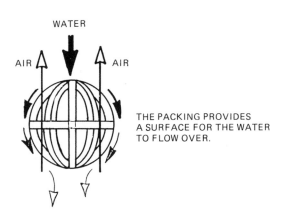

WATER

AIR   AIR

THE PACKING PROVIDES
A SURFACE FOR THE WATER
TO FLOW OVER.

THE CONTAMINANT IS DRIVEN OUT
OF THE WATER BECAUSE OF THE
DIFFERENCE IN CONCENTRATION
BETWEEN THE AIR AND WATER.

FIGURE 3-14. Tower packing.

TABLE 3-8   Characteristics of Several Polypropylene Packings

Packing*	Nominal Size	Weight lbs/ft³	Surface ft²/ft³	Free Vol %	Packing Factor per ft
Tripacks	1″	6.2	85	90	28
Tripacks	2″	4.2	48	93	16
Tripacks	3.5″	3.3	38	95	14
Berl Saddles (ceramic)**	1″	45	79	69	NA
Pall Rings	2″	3.85	33	92	25
Novalox-Saddles	1″	5.2	78	81	33
Tellerettes (No. 3 Typol)	3.75″	4.7	30	92	NA
Hi-Flows	2″	3.7	33.5	94	NA
Lampacs	3.5″	4.2	45	93	14
Munters 12060 Structural	–	NA	68	95	27.4

NA = Not available.
*Registered trademarks. Source: Manufacturer's literature.
**Source: *Chemical Engineer's Handbook*, 4th edition.

One of the great improvements in mass transfer technology has been the introduction of inexpensive plastic packings. The use of injection molding has allowed the creation of packings much more suited to the dual goals of maximum mass transfer surface and minimum pressure drop. Early packings, such as berl saddles and rashig rings, were limited to fairly simple shapes by the nature of their production process, usually in metals or ceramics. The packings shown in Figure 3-15 are all made of polypropylene, and their complex shapes assure a large void area to minimize pressure drop. Pilot testing of these packings have shown much better results than those obtained with the older packings. Polypropylene packing has several other benefits. It is chemically inert, and will not degrade when exposed to most chemicals encountered in groundwater contamination. (Extreme levels, greater than 1000 ppm, of certain chlorinated compounds, particularly methylene chloride, can lead to a softening of polypropylene and subsequent degradation of the mass transfer process.) Polypropylene is very inexpensive; an equal volume of ceramic packing may cost 8- to 10 times as much. Finally, it is lightweight and strong, allowing greater packed bed depths without crushing the packing at the base of the tower. This strength also allows the packing to be dumped into a tower without damage; ceramic packing must be loaded into a tower filled with water to cushion its fall. Designing a tower to withstand such hydraulic loading adds to the cost of installation.

FIGURE 3-15. Examples of plastic tower packing. (Courtesy of Hydro Group.)

In situations where polypropylene is not appropriate, more resilient plastics such as Kynan™ or Teflon™ are available, but at prices 14 to 40 times that of polypropylene. Stainless steel packings offer a more cost effective alternative to Kynan or Teflon in most extreme situations, although packings made of stainless steel material are generally less efficient than their plastic counterparts.

The packings shown in Figure 3-15 all fall into one general category: the random dumped packings. These packings are simply dumped into the tower and allowed to rest in whatever configuration they land. The other broad category of tower packing is the structured packings.

These packings are physically stacked in the tower, and configured in such a way as to assure even water redistribution. These packings are not generally used in packed column air stripping due to their high capital cost, the additional labor cost of their installation, and somewhat poorer mass transfer efficiency. They exhibit lower headloss characteristics, thus higher air to water ratios are achievable for the same power cost. Claims of superior fouling resistance have been made; however, these claims have not been verified in field operations.

The final component of an air striping system is the air delivery system. Usually a forced draft blower is provided at the base of the tower, or housed nearby in a building if sound levels are a concern. However, with proper tower design, selection of a minimum air to water ratio and minimization of the fan speed; the size and noise level of a blower can be kept to a minimum. Sound mufflers are available for insertion over the air inlet if desired, but these result in an increased pressure drop and only marginal reduction of the noise level. On some potable water installations, air filters are required to ensure no air particulates contact the water. These filters also serve as excellent sound mufflers. A complete packed tower is shown in Figure 3-16.

The alternative to a forced draft tower is an induced draft system

FIGURE 3-16. Example of a packed tower air stripper. (Courtesy of Hydro Group.)

where air is drawn through the tower by the blower. The blower may be mounted on top of the tower or at the base. Induced draft units with blowers mounted on top are limited to somewhat lower pressure drops. Base mounted induced draft systems often find application where the gases being discharged undergo further treatment before their release, i.e., carbon adsorption or incineration.

Operation

Once installed, the operation of an air stripper can be a very simple matter. The air delivery system is controlled by the water supply pumps, so that before water is introduced to the tower, the blowers are activated. Since the only moving part is the blower, maintenance of the mechanical system is fairly straightforward. Thus, maintenance can be minimal, only requiring periodic inspection of the packing. However, under some circumstances maintenance can be a major consideration in tower operation. There are several considerations that can increase required maintenance.

Overall system maintenance will vary with several factors: materials of construction, environmental conditions, and water quality. The use of carbon steel, for instance, will require regular recoating to maintain the carbon steel shell or tower internals. FRP must also be maintained by recoating any exposed FRP components to prevent UV degradation.

The environment can also contribute to maintenance work. Leaves and snow have been reported to clog air intake structures, and improperly designed systems have led to frozen pipelines. Freezing can be avoided by designing the tower and internals to be entirely self-draining, avoiding any interior ledges or pockets that could develop more than a thin coating of ice. Inlet and outlet piping should be provided with normally open solenoids that are heat traced and insulated to allow drainage of exposed pipelines upon shutdown.

During normal operation, freezing is not a problem within the air stripping tower except where the inlet water temperature drops below 35°F or in extremely cold climates (Alaska, for example).

Water Quality Considerations
Certain inorganic water quality parameters pose operation and maintenance concerns when considering the use of packed column aera-

tion systems. Of particular note are the effects of dissolved iron, suspended solids, high microbial populations (degradable organics), and hardness.

During the aeration process, dissolved metals such as iron and manganese are oxidized. In most situations, the pH of the water is such that manganese deposition is not a problem, however the transformation of ferrous iron to ferric iron is a real and often destructive problem in air stripping operations. The oxidized iron deposits on the packing material in time cause a build up that will bridge and clog the packed bed; this leads to a decline in system efficiency. The degree to which iron deposition effects system performance is directly related to the level of dissolved iron in the water and other water quality characteristics such as pH and dissolved solids. Iron deposition can be controlled through the application of pretreatment technologies such as aeration/filtration and chemical sequestering. The effectiveness of any control technology will be related to the level of iron present.

Iron deposition is probably the most underestimated problem with groundwater treatment systems in general, and air strippers specifically. The problem starts when iron is not considered a contaminant and is not included in the analyses during the remedial investigation. The designer needs to know the concentration of iron, manganese, hardness, and other inorganic constituents in the water to correctly design an air stripper. Depending on the iron concentration, the packing may have to be cleaned as often as every month. As a very rough rule of thumb the following guidelines can be used:

<1 mg/l Fe—Low maintenance
2-5 mg/l Fe—Clean tower every 3-6 months
>10 mg/l Fe—Clean tower every 1-2 months

All of these values are also affected by temperature, pH, manganese, hardness, and other environmental factors. For example, the lower the pH, the slower the iron oxidation reaction. If the pH of the groundwater is 4.5 to 5.0, then iron precipitation will be greatly reduced as a maintenance problem. These concentration ranges should only be used as a rough guide line.

The added cost of maintenance and the lower removal efficiency, due to fouling may be sufficient to force pretreatment or selection of alternative air stripper designs. Pretreatment systems can cost as much as if not more than the actual air stripper. The iron itself is easy to remove from water. Where oxygen is a problem in the tower it can also be used as an opportunity to easily remove the iron from the water. If the groundwater is aerated prior to discharge to the air stripper, then the iron will come out of solution and form insoluble iron hydroxide. This reaction occurs at a neutral pH (see Chapter 5, Inorganic Treatment, for detailed information on iron solubility at different pHs).

Once the iron is in solid form, then the problem is solids/liquid separation. The iron particles are removed from the water by a clarifier or filter. Usually, the suspended solids concentration is low enough for a filter to be used. Once the solids have been removed, then the groundwater can continue to the packed tower. The solids require more processing. Basically, the solids must be reduced to as small a volume as possible for final disposal (probably in a landfill). The solids will have to be sent to a thickener, and then to some type of dewatering device (filter press, belt press, etc.).

As can be seen, several unit operations are required to remove the iron from the water. There are added capital costs for all of the extra units and the operation cost will increase for chemicals, manpower, electricity, and sludge disposal. On large scale systems, these costs will probably be more than the cost of the air stripper. On a small scale system, the designer will probably find the costs for pretreatment prohibitive, and switch to a different air stripper design. For example, a diffused aeration stripper can be used with high iron concentrations with no adverse affect from the iron. This technology will be covered in detail under the Alternative Methods section of this chapter.

The presence of high populations of microbial bacteria and/or high concentrations of degradable organics can lead to a biological build up within the packed bed. This problem occurs because the packing material and highly oxygenated water offer an excellent environment for microbial growth. As with oxidized iron, a biological build up can lead to a deterioration of system performance. Biological build ups are relatively uncommon in packed column systems treating groundwater for municipal drinking water applications. The problem occurs more often in situations involving groundwater clean-ups of petroleum spills, landfill leachate treatment or any time there are higher (>10 mg/l) concentrations of degradable organics. This problem is controlled through the use of chlorine (or other oxidizing agent) solutions generally added prior to the air stripper on an intermittent or constant basis. One note of caution when adding oxidizing agents to packed towers. The oxidizing agents can also increase the rate of oxidation of the inorganic constituents of the groundwater. For example, solving a biological buildup with chlorine oxidation can cause manganese to precipitate when it normally stays in solution.

One solution to bacterial buildup is to pretreat with a biological treatment unit. The biological unit can remove the degradable organics, and then the air stripper can remove the nondegradable organics. Low concentration biological treatment designs are discussed in the next chapter.

Calcium hardness is another operation/maintenance consideration. In the air stripping process, the potential for destabilization of the water is increased as a result of the removal of dissolved carbon dioxide from the water. The removal of carbon dioxide can lead to calcium carbonate deposition within the packed tower and in any post treatment distribution system. If the destabilization is sufficient and occurs over a long enough period of time the packing will become clogged with calcium carbonate. The problem can be controlled by minimizing the air to water ratio as much as possible (thus minimizing the stripping effect), the use of carbon dioxide, or through pretreatment. Pretreatment can take the form of softening systems or chemical sequestering agents.

More packed tower air strippers have failed because of maintenance problems than from bad designs. While most of this section has covered the details of the packed tower design, the real world requires a complete understanding of the environmental factors that will affect the packing and subsequent operation of the tower.

Air Treatment

One of the major concerns about air stripping is the discharge of volatile organics into the atmosphere. The contamination is not destroyed in a mass transfer process; it is merely transported into another medium. Two factors mitigate the effects of these atmospheric discharges. The first factor is the dilution that takes place in the tower before the vapors are emitted. Air to water ratios commonly employed range from 25:1 to 250:1. Thus, the pollutant is diluted by a similar factor when it is transferred into the air. In addition to this dilution, there is a natural dilution that occurs as soon as the air stream is dispersed into the atmosphere. The second factor is that many compounds, such as TCE and PCE, will break down in the atmosphere under the effects of the sun's radiation. TCE, for example, has a half-life of approximately a day and a half in the atmosphere. Environmental treatment in the U.S., however, has tended toward destruction or final disposal, not toward switching the pollutant from one medium to another.

Already several states require that all air discharges from stripping towers be treated before being released to the atmosphere. In these states and in circumstances where the total discharge to the atmosphere is too high, the exhaust gases are usually treated using one of these means: activated carbon, incineration, or chemical destruction. The first two methods have proven to be the most viable approaches and are readily available in the marketplace; both are discussed below.

Activated Carbon

By far the most common control technology applied to air strippers is vapor phase granular activated carbon (VPGAC). The adsorption process in the vapor state is similar to the process as it occurs in the liquid phase (see Carbon Adsorption section in this chapter), the major difference being the fluid treated. Detailed discussions of the adsorption mechanism and its application to air stripper offgas treatment are offered in several of the references listed.[14,15,16]

At first glance vapor phase carbon appears superfluous, since liquid phase carbon could treat the water directly. However, this system may save on carbon costs, because the mass transfer in vapor phase carbon is much faster so that smaller beds can be used and carbon usage decreased. In addition, the pore size distribution in granular activated carbon, manufactured specifically for vapor phase treatment, allows for more of the surface area to be used for adsorption and for a greater capacity for adsorption of chemicals in the vapor phase as compared to the liquid phase. There will also be fewer chemicals in the vapor stream competing for the available pore space, since nonvolatile compounds if present, will remain in the liquid phase.

VPGAC can be employed in systems which provide for off site disposal or regeneration of the activated carbon or in systems that utilize on-site regeneration facilities. The nonregenerable systems are much more common in air stripping applications due to the comparative low capital cost associated with the technology and the relatively low levels of VOCs present in most air stripping applications, See Figure 3-17. The biggest drawback of the nonregenerable systems is the need to remove, dispose of, and replace the GAC, typically as hazardous waste, on a regular basis. Several companies offer regeneration services on a commercial basis.

Regenerable VPGAC processes rely on in-place regeneration of the activated carbon to reestablish at least part of the carbon's adsorptive capacity. Steam is usually used for regeneration. Other means, such as hot gas, have become commercially available, and can be combined with thermal oxidation of the hot regenerant gas to provide for complete organic contaminant destruction on-site. The use of steam regenerable VPGAC systems will be examined here, as the process design for other regenerants is very similar.

The typical regenerable VPGAC system has two modes of operation: adsorption and desorption. During the adsorption step, the air to be treated is passed through the activated carbon bed. There are

FIGURE 3-17. Packed tower air stripper with VPGAC.

many different types of bed configurations: thin bed, cartridge, and deep bed are common. The configuration which is chosen is based on the compounds treated and the design efficiency required.

Preheating of the air is almost always required with air stripping operations. First, we must insure that the water in the air stream does not come out of vapor phase and deposit on the carbon granules. Adsorption, considering both capacity and mass transfer, improves substantially as relative humidity is lowered to about 40%, after which little improvement is gained (this also holds true in nonregenerable systems). Unlike nonregenerable VPGAC, once breakthrough has occurred, the carbon is not replaced. With a regenerable VPGAC process, the air treatment system is taken off line and the carbon is regenerated in place. Often dual column systems are designed to insure that air treatment can continue even during the regeneration step. Since breakthrough will vary with the actual organic load, air humidity and temperature, and carbon age, timers are often used to take the VPGAC unit off line on a regular basis. The timing of the regeneration process may be changed as the system ages. The steam regeneration is much less rigorous than higher temperature regeneration, so the actual, or "working" capacity will be less for virgin grade ore regenerated carbon, so cycle times will be much shorter than for a similarly sized nonregenerable VPGAC system (hot air will provide greater working capacities than steam).

The regeneration step consists of flowing steam in a counter-current pattern through the carbon bed. The steam will desorb or remove the sorbed organic molecules from the carbon pores and carry them out of the carbon bed. This desorption process will typically last about an hour. All the steam used in the desorption is run through a condenser. The condensate and free product resulting from the desorption process are collected. In some cases, the free product is then reused, in other cases it is disposed of. The aqueous condensate is saturated with the organics removed from the carbon. When the pure product is removed, the aqueous condensate may be recycled slowly back to the packed tower for blending and treatment with the normal influent. The aqueous condensate may need to be disposed of after a number of cycles, if the concentration of the more soluble contaminants becomes too great to blend with the influent.

Once the desorption step is complete, a cool-down period may follow, after which the regenerated unit is put back on line. As the

VPGAC is regenerated over and over again, it loses some of its original capacity. Generally the expected life of the VPGAC is approximately five years before it must be replaced. VPGAC units running on single compounds will have long carbon lives, but when the carbon is subjected to a mixture of organics, the carbon bed life can be reduced significantly.

Incineration

Incineration of the air stripper off gas is a second means of treatment. Direct thermal incineration is sometimes used in air stripping applications for landfills where methane flares are commonly in use; the stripper off gas is directed to the flares and incinerated at temperatures near 1400°F. Other incinerators will run between 1400°F and 1800°F. Catalytic incineration utilizes a catalyst to enhance the destruction of the organic compounds. A catalytic system can operate at reduced temperatures, significantly contributing to operating cost savings associated with fuel costs. Catalytic incinerators are limited to applications with nonchlorinated hydrocarbons. The chlorine molecule fouls the catalyst. Therefore, catalytic systems are usually only used on petroleum hydrocarbons. Some new types of catalysts claim to work with chlorinated hydrocarbons. It is suggested that long-term pilot plants be run on these applications until several full-scale installations are designed and operated on compounds similar to the contaminants at a particular remediation.

All the air discharge control processes discussed above are commercially available in the United States from several equipment suppliers. The addition of the simplest air treatment system will approximately double the capital cost of an air stripping system. The use of on-site regeneration or incineration will increase the cost of treatment by as much as one to two dollars per thousand gallons treated. These are very general cost estimates that are strongly influenced by the degree of treatment required, the compounds, and levels of organics present, location of the site, and regulatory requirements.

Alternative Air Stripping Methods

There are a wide variety of aeration methods available in the market today. Some are listed in Table 3-4. Of the technologies listed, packed columns have found the most widespread use. Slot tray aerators provide low treatment efficiencies but are not subject to the same fouling problems that plague packed towers. Cascade aerators have no operating cost associated with the air supply, since they are open to the atmosphere. They simply do not provide adequate air/water contact for most groundwater remediations. Removal efficiencies obtained using cascade aerators are much less than those obtained with packed towers.

Diffused air or bubble aeration air strippers have probably seen the largest increase in application outside of packed towers during the last five years. These systems are mainly applied to small flows, less than 50 gpm, and in situations that have high iron content or other material that may foul the packing of a packed tower air stripper. Basically, designers have been finding that the maintenance cost of a packed tower air stripper is the controlling design parameter in small systems. If iron is present in the water then a low maintenance system like a diffused aeration system is preferable.

These systems can also be designed to be compact and portable, and can be shaped to completely fit into small buildings at a treatment site. Certain designs, D2M2, Figure 3-18 also add filtration directly in

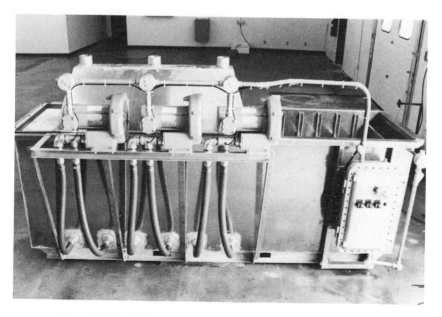

FIGURE 3-18. D2M2 air stripper. (Courtesy of Geraghty & Miller, Inc.)

the unit to produce high quality effluent from the air stripper. This technology has been widely applied to gas station clean-ups because of its ease of operation and flexibility which allow the technology to be applied over a wide range of low flow conditions.

Innovative manufacturers have devised methods of cascading the water being treated through a series of aeration chambers to increase efficiency of the bubble aeration process. These systems offer a low profile while taking up a relatively small amount of space.

Spray basins have found use in several groundwater cleanup cases. In these systems, a piping grid is laid out over the area of a basin (either earthwork or concrete) and spray nozzles are placed evenly throughout the area to spray the contaminated water into the air in very fine droplets. The water is then either collected in the basin and pumped off, or, in the case of an earthwork recharge basin, allowed to percolate back into the ground. This method is often used to flush out an area of contaminated soil. The advantage of this system is its extremely low capital cost, which makes it ideal for a cleanup of a temporary nature. The disadvantages are that large tracts of land are required, and that neighboring properties may be affected by wind-driven mists or, in the winter, ice crystals. One installation in the northern U.S. solved this problem by erecting a 40 foot high tarp along a property line, but this is a stopgap solution at best. Extra pumping costs may also be incurred to provide adequate pressure to the nozzles. Finally, regulatory concerns will have to be satisfied when spraying hazardous waste onto land.

One of the newest designs for air stripping is the rotary stripper. In this unit, the water is passed through a chamber that is rotating at high speeds. The rotary stripper has found limited application in situations where a medium efficiency, low profile system is required. The power costs associated with rotary strippers are extremely high when compared to packed towers, making them an expensive alternative.

A relatively new approach for VOC removal employs the tray technologies which have been used in distillation columns for decades. Air is forced in below an orifice tray which contains a flowing water stream. The airflow is designed to maintain a "froth" in the tray to provide the mass transfer required. Trays can be stacked and run in series in order to increase system performance, or water flowrate can be reduced over a single tray to enhance removal efficiency. Since there is no packing, the potential for fouling is reduced. These systems require a great deal of energy and are not cost effective above flows of approximately 100 gpm.

There are several new miscellaneous stripping methods. The use of cyclone aeration systems, hollow fiber membrane air strippers,[17] and hollow fiber membrane/oil stripping[18] are being explored as alternatives to packed column aerators.

It is important to remember, when designing a treatment system, that there are other types of air strippers. Do not simply put a Band-Aid™ on a packed tower design. Sometimes, go back and start from the beginning. If maintenance is going to be a problem, evaluate alternative designs.

Air stripping has become one of the workhorse technologies for the treatment of VOCs in groundwater. The combination of low cost, easy operation, and the wide variety of compounds that can be removed from groundwater make stripping the first choice for many low concentration streams. Even in cases when the air discharge must be treated, air stripping is still often found to be the least-cost method.

CARBON ADSORPTION

The use of carbon for its adsorptive qualities was known as early as 1550 B.C., when charcoal was utilized in the purification of medicines. In the field of water treatment, both the ancient Egyptians and eighteenth century sailors used charcoal lined vessels to provide for clean drinking water.

The use of carbon in a water treatment process, though, traces its roots to London in the 1860s, when some of the residents of that city had their drinking water filtered through animal charcoal to remove tastes and color. Granular carbon filters were introduced in the 1930s for producing ultrapure water for the food and beverage industry.

Following World War II, coal was used to produce high activity, hard granular carbons on a commercial scale, leading to the widespread use that granular activated carbon has today.

The use of activated carbon to remove taste and odors from drinking water supplies is now an established technology. Since the introduction of activated carbon on a commercial scale, industry has also taken advantage of the material's unique ability to adsorb a variety of organic compounds for product purification, water and wastewater treatment.

Based on its history and unique properties, activated carbon has now become a proven technology for removal of synthetic organic contaminants from groundwater. Though these contaminants do not exhibit traditional taste and odor characteristics, and may be present in trace level concentrations rather than the high levels found in wastewaters, removal may be required because of toxicity concerns.

Concepts of Adsorption

Adsorption is a natural process in which molecules of a liquid or gas are attracted and then held at the surface of a solid. Physical adsorption refers to the attraction caused by the surface tension of a solid that causes molecules to be held at the surface of the solid. Chemical adsorption involves actual chemical bonding at the solid's surface. Physical adsorption is reversible if sufficient energy is added to overcome the surface's attractive forces, while chemical adsorption is not a reversible reaction.

Adsorption on activated carbon is of a physical nature. What makes this material such an excellent adsorbent is the large amount of surface area that is accessible for the adsorption process contained within the carbon particle. The surface area of granular carbons range up to 1400 square meters per gram of material. As the surface area of the activated carbon is internal to the material, crushing the granular material will not increase its surface area. Even in its crushed or powdered state, activated carbon still retains its internal surface area, making it an effective adsorbent.

Two methods have been developed to describe the internal structure of the carbon particle. The original method used macropores and micropores as a basis of description. Figure 3-19 shows macropores (openings with diameters greater than 1000 A) where the organics are transported to the interior of the carbon particle, and micropores where most of the adsorption takes place.

Figure 3-20 is a micrograph of an actual carbon particle. In this picture, the carbon particle has transport pores, or openings that allow passage of the contaminant molecules and provide an entrance into the interior structure of the carbon particle. Although some adsorption may take place along these pores, they mainly serve to conduct the molecules to the adsorption pores where adsorption takes place. Many such adsorption pores are large enough only to contain small molecules, so the effective surface area for adsorption

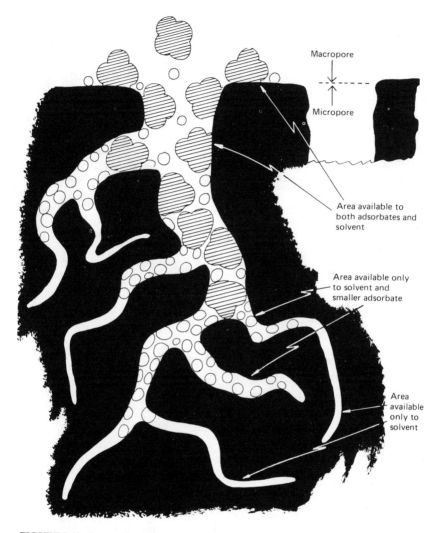

Macropore

Micropore

Area available to both adsorbates and solvent

Area available only to solvent and smaller adsorbate

Area available only to solvent

FIGURE 3-19. Internal structure of activated carbon. (Courtesy of Calgon Carbon Corp.)

of a particular species depends upon its size and the available surface area of the pores it can enter.

In either case the capacity of a particular grade of carbon, may vary for different species. Standard tests have been developed to identify such capacities. These tests may utilize iodine molecules to identify small pores, and molasses to identify the larger pore structure, for example. There are a wide variety of activated carbons available,

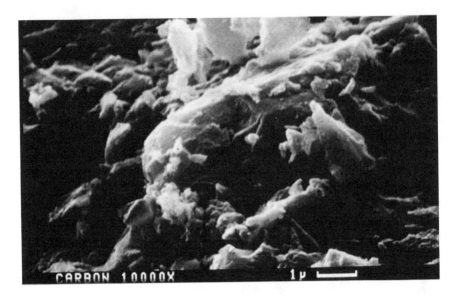

FIGURE 3-20. Micrograph of an activated carbon particle. (Courtesy of Calgon Carbon Corp.)

and properties such as surface area, and pore size distribution will determine their applicability to any given situation.

The adsorption mechanism consists of three steps: (1) diffusion of the molecules through the liquid phase to the carbon particle, (2) diffusion of the molecules through the transport pores (macropores) to the adsorption site, and (3) the adsorption of the molecule to the surface. The characteristics of the molecule will determine the rate of each step and finally the amount of time required for the entire adsorption process. Less soluble organics, for example, will diffuse rapidly to the granule, and large molecules will move slowly through the pore structure. Generally, the chlorinated solvents found to be contaminating groundwater are amenable to activated carbon adsorption, due to their low solubility and small molecular size which enables effective use of the adsorption area in smaller pores. However, it is important for each contamination problem to be properly evaluated.

Evaluation Procedures—Adsorption Isotherms

The first step in evaluating activated carbon adsorption for a specific application is to assess its feasibility utilizing a liquid phase adsorption isotherm test.

An adsorption isotherm test is a batch test designed to demonstrate the degree to which a particular dissolved organic compound (adsorbate) is adsorbed on activated carbon (adsorbent). The data generated shows the distribution of adsorbate between the adsorbent and solution phases at various adsorbate concentrations. From the data, a plot of the amount of impurity remaining in solution at constant temperature can be generated. For a single adsorbate, a straight line plot (on log - log paper) can be obtained when using the empirical Freundlich equation:

$$x/m = kc^{1/n} \text{ or } \log x/m = \log k + 1/n \log c \qquad (3\text{-}5)$$

where:

x = The amount of contaminant adsorbed
m = Weight of carbon
c = Equilibrium concentration in solution after desorption
   k and n are constants

For mixtures of adsorbates, a series of straight lines can be obtained. The presence of a nonadsorbable component will result in a curvature of the line, when in combination with an adsorbable component, and in a vertical line when alone.

Data for generating this type of isotherm are obtained by treating fixed volumes of the water sample with a series of known weights of carbon. The carbon-liquid mixture is agitated for a fixed time at a constant temperature. After the carbon has been removed by filtration, the residual adsorbate concentration is determined. The amount of organic adsorbed by the carbon (x) is divided by the weight of carbon in the sample (m) to give one value of x/m for the isotherm.

For contaminants that are volatile at ambient temperatures, the isotherm tests are conducted utilizing water samples with no head space, to prevent loss of contaminants to volatilization.

To estimate the capacity of carbon for the contaminant then, one uses the x/m value that corresponds to its influent concentration C_o. This value of (x/m) represents the maximum amount of contaminant adsorbed per unit weight of carbon when the carbon is in equilibrium with the untreated contaminant concentration. Table 3-9 presents the

TABLE 3-9   Carbon Adsorption Isotherm for Trichloroetylene

m Carbon (g)	c, TCE Remaining (ppm)	(mg)	x TCE Adsorbed (mg)	x/m
Control	1.600	0.800	—	—
0.0005	1.490	0.745	0.055	110.0
0.0010	1.520	0.760	0.040	40.0
0.0025	1.290	0.645	0.155	62.0
0.0050	1.060	0.530	0.270	54.0
0.010	0.860	0.430	0.370	37.0
0.025	0.285	0.143	0.657	26.3
0.050	0.165	0.083	0.717	14.3
0.100	0.035	0.018	0.782	7.8
0.250	< 0.010	—	0.800	—
0.500	< 0.010	—	0.800	—

Conditions:
Type of carbon — Filtrasorb 300
Temperature   — ambient
Sample volume — 500 ml
Agitation time — 4.0 hr

isotherm data for a TCE at 1600 μg/l. The data are summarized in Figure 3-21.

As an example, assume TCE was present in groundwater at 200 μg/l. According to the isotherm, the equilibrium capacity is 19.5 mg trichloroethylene adsorbed per gram of carbon, or a capacity of about 2%. Therefore, the amount of carbon required would be: (0.2 mg/l)/(19.5 mg/g) = 0.01 gram carbon/liter of water, or approximately 0.1 lbs per 1000 gallons treated. This capacity is based on allowing the activated carbon to reach equilibrium with TCE, an ideal condition not usually obtainable in practice. But, the isotherm evaluation does prove that carbon adsorption is feasible and should be evaluated further.

Table 3-10 shows the equilibrium adsorption capacities of some typical groundwater contaminants (synthetic organic solvents and other compounds) as determined from isotherm testing.

Evaluation Procedures—Dynamic Column Study

To design an activated carbon adsorption system, additional information that is not available from the adsorption isotherm must be obtained. The optimum operating capacity and contact time need to

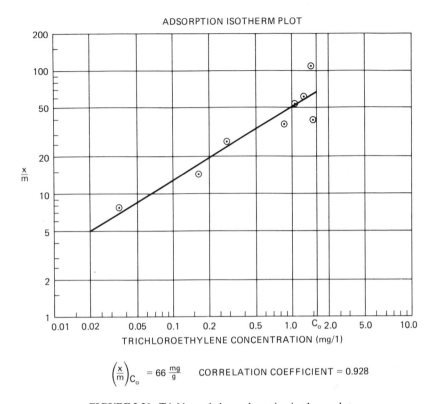

ADSORPTION ISOTHERM PLOT

$$\left(\frac{x}{m}\right)_{C_o} = 66 \frac{mg}{g} \qquad \text{CORRELATION COEFFICIENT} = 0.928$$

FIGURE 3-21. Trichloroethylene adsorption isotherm plot.

be determined to fix the adsorber size and optimum system configuration. The optimum contact time and mass transfer zone depend upon the rate at which the contaminant is adsorbed by the carbon, and can only be determined by dynamic testing.

The column test is conducted with a series of columns connected in series, as shown in Figure 3-22. Each column is filled with an amount of carbon which has been calculated to provide superficial contact times of 15 to 60 minutes per column. The liquid rate to the column is usually in the range of 2 gpm per square foot, although it may vary during the test, because at this point of the evaluation the contact time is of more importance. The surface loading rate may be of more importance if there are suspended solids present and the activated carbon bed is to act as a filtering medium as well as an adsorption material.

Water is pumped through the column system and effluent samples

TABLE 3-10   Adsorption Capacity for Specific Organic Compounds

	Compound	Adsorption Capacity (mg compound/g carbon) at 500 ppb	Reference
1	Acetone	43	1
2	Benzene	80	1
3	Bromodichloromethane	5	4
4	Bromoform	13.6	4
5	Carbon tetrachloride	6.2	2
6	Chlorobenzene	45	3
7	Chloroform	1.6	1
8	2-Chlorophenol	38	3
9	p-Dichlorobenzene (1,4)	87.3	4
10	1,1-Dichloroethane	1.2	4
11	1,2-Dichloroethane	2	2
12	1,1-Dichloroethylene	3.4	4
13	cis-1,2-Dichloroethylene	9	5
14	trans-1,2-Dichloroethylene	2.2	4
15	Ethylbenzene	18	1
16	Hexachlorobenzene	42	3
17	Methylene chloride	0.8	3
18	Methylethylketone	94	1
19	Methyl naphthalene	150	5
20	Methyl tert-butyl-ether	6.5	5
21	Naphthalene	5.6	3
22	Pentachlorophenol	100	3
23	Phenol	161	1
24	Tetrachloroethylene	34.5	2
25	Toluene	50	1
26	1,1,1-Trichloroethane	2	2
27	1,1,2-Trichloroethane	3.7	4
28	Trichloroethylene	18.2	2
29	Vinyl chloride	Trace	3
30	o-Xylene	75	4

1. Verschuren, Karel. *Handbook of Environmental Data on Organic Chemicals.* New York: Van Nostrand Reinhold, 1983.
2. Uhler, R. E., et al. *Treatment Alternatives for Groundwater Contamination.* James M. Montgomery, Consulting Engineers.
3. Stenzel, Mark. Letter of Correspondence to Evan Nyer, August 22, 1984.
4. USEPA, *Carbon Adsorption Isotherms for Toxic Organics,* EPA-600/8-80-023, Municipal Environmental Research Laboratory, April 1980.
5. Roy, Al. *Calgon Carbon,* (personal correspondance) 1991.

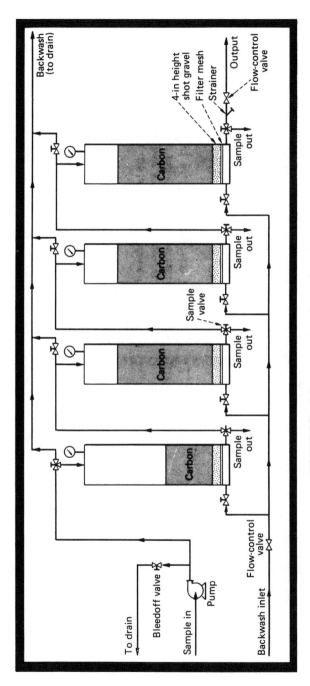

FIGURE 3-22. Laboratory series column adsorption test.

are collected from each of the columns. The adsorption isotherm test should provide an estimate of how often testing should be done. The amount of the contaminant in the column effluent is plotted against the volume throughput of each column. The result is a series of curves, each curve representing a column. The successive curves also represent increasing contact times in a single bed. Figure 3-23 shows an example of a column study where each column represents 15 minutes of contact time. The curves obtained are termed break-through curves, as they represent the concentration or amount of contaminants present in the effluent (which have passed through the column unabsorbed).

The results of a dynamic column study are used to establish the design of an operating carbon adsorption system. The first step is to establish the contact time required in the operating system. For each of the breakthrough curves established in the column study, a carbon usage rate can be calculated. This usage rate determines the pounds of carbon required for a given volume of liquid to maintain the contaminant at a desired level in the effluent. The usage rate is calculated by dividing the amount of carbon on-line by the volume of water treated when the desired effluent concentration is exceeded, or the breakpoint of the breakthrough curve. The carbon usage rates

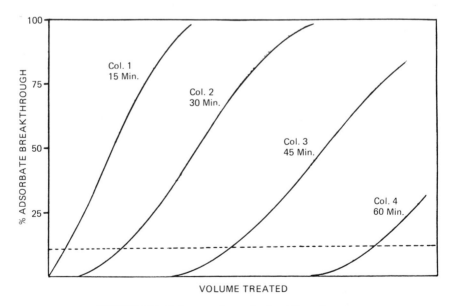

FIGURE 3-23. Column study results: breakthrough curves.

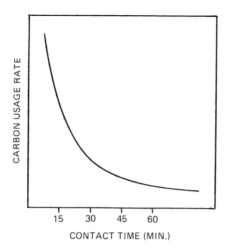

FIGURE 3-24. Optimum carbon contact time.

can then be plotted for each contact time (column) as shown in Figure 3-24, and the optimum contact time determined as the point where increasing contact time obtains little improvement in carbon usage. The amount of carbon on-line is then established by multiplying the contact time by the flow rate to obtain the volume of the carbon bed.

The next step is to consider whether only a single carbon adsorber is required, or if a second unit in series would yield substantial benefits. Figure 3-25 shows the configuration of two breakthrough curves. The steep curve indicates a relatively short mass transfer zone. In this case, good utilization of the activated carbon can be realized in a single bed where the carbon is exchanged when the effluent concentration exceeds the desired level. The gradually sloping curve indicates a long mass transfer zone. For these instances, a staged system would provide for more optimum usage. A second stage will maintain final effluent quality while the effluent from the first step gradually rises to near influent concentrations, utilizing all of the adsorptive capacity of the carbon. When the carbon in the first unit is fully utilized, it is then replaced with fresh carbon, and put back on-line as the final stage, allowing full utilization of the carbon in the other unit now in service as the first stage. The concept of staged adsorbers is of particular value when considering a water requiring treatment which contains a variety of contaminants exhibiting differing adsorbities.

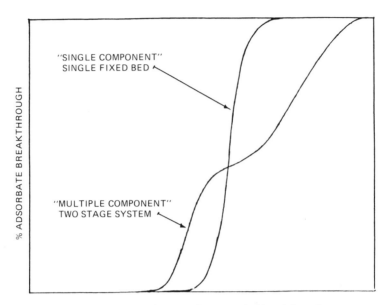

FIGURE 3-25. Mass transfer zones for two typical breakthrough curves.

After the contact time has been established, and the evaluation of the breakthrough curves has indicated whether a single bed or a staged system is preferred, the designer can select the adsorber configuration. If the breakthrough curve is steep, usually in the case of single or similar contaminants, the single fixed bed downflow adsorber is the most economical choice. The contact time will establish the total carbon volume as noted above. By weighing considerations such as flow and carbon volume, the designer will select the vessel size and whether multiple units (operated in parallel) may be required.

The simplest downflow fixed bed is the gravity adsorber. As the downflow unit operates as a filter, and is not a pressure vessel, flow is limited to 2 gpm/ft². At 2 gpm/ft² the typical carbon depth is 4 feet, yielding a contact time of 15 minutes. The effluent from a gravity adsorber will require pumping if needed for a pressure water system. As the system depends upon gravity as the motive force, the gravity adsorber may require backwashing if suspended solids are present in the influent.

The fixed bed system can also be contained in a pressure vessel. This vessel allows greater bed depth, and higher surface loading rates

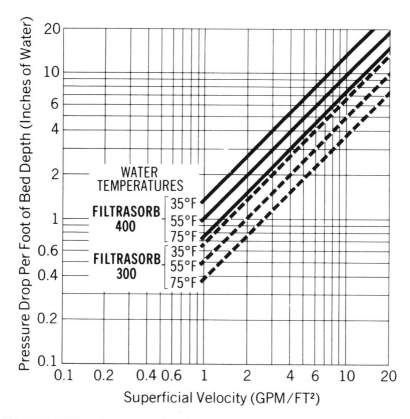

FIGURE 3-26. Downflow pressure drop through a backwashed segregated bed of Filtrasorb 300 and Filtrasorb 400. (Courtesy of Calgon Carbon Corp.)

(up to 5 gpm/ft^2), or greater contact times. This system can also be operated at higher pressures, so the unit could be placed in-line between the pump and downstream usage. The pressure drop through a typical granular carbon bed is shown in Figure 3-26.

If the breakthrough curve is gradual or discontinuous (has a temporary plateau value) due to multiple contaminants, the designer may wish to specify a staged carbon adsorption system to obtain more optimum utilization of the adsorptive capacity of the carbon. The simplest staged adsorption system is two single fixed beds in series. If the mass transfer zone can be maintained within a single bed, then the second stage will be able to maintain effluent quality, while the carbon in the first stage is obtaining full use of its adsorptive capacity. When the carbon in the first stage is fully utilized, it is exchanged for fresh carbon and returned to service as the second stage. The fixed bed downflow system has the added advantage of operating as a media filter with elimination of suspended solids in the effluent.

Another form of the staged bed system is the upflow moving bed design. This system may be of use when long contact times are required and the breakthrough curve indicates that even a two stage system is insufficient to provide economical use of the carbon.

In this process, the carbon is placed in a large coned bottom vessel. The cone bottom is desirable, as removal of carbon will be of the mass (plug) flow variety in which material at the side walls will move downward in the vessel at the same rate as that in the center. Flow from the center only is termed rat-holing, and results in an uneven distribution of fresh and partially spent carbon in the bed, and may cause premature breakthrough.

The water flow is directed up through the bed at rates up to 8 gpm/ft^2. At this rate, there may be a slight expansion of the bed. This will produce a small amount of carbon fines material in the effluent, which may constitute a problem requiring nominal filtration.

When a portion of the carbon is fully utilized, it is withdrawn from the bottom of the unit, and an equal volume of fresh carbon is placed at the top of the bed. The portion of the bed requiring replacement may be anywhere from 5 to 50%, depending upon the breakthrough curve. Replacement volumes of 50% or greater usually indicate that a two-staged fixed bed system may be a better selection. It is very unusual to find an upflow carbon adsorber used in groundwater cleanups.

Granular Activated Carbon Replacement Considerations

The supply of granular activated carbon to the adsorption system may be a significant operating cost factor. Usually the usage rate in a groundwater treatment system is not at the level which would justify consideration of an on-site regeneration facility. The most common form of replacement is to recharge the unit with fresh activated carbon, which can be virgin carbon or reactivated carbon, with the reactivation occurring off-site.

As is often the case in groundwater treatment, the contaminant may be a volatile organic solvent. If there are no nonvolatile contaminants present, recent studies have shown that much of the adsorptive capacity may be recovered by regenerating the carbon in place, or in situ. This can be accomplished in a fixed bed by withholding the unit from the process and passing steam through the bed. The regeneration in a moving bed can be effected by removing the carbon from the bed, exposing it to steam in a separate unit, and utilizing the regenerated material as fresh material at the top of the bed. Depending upon the process, carbon can be utilized from 5 to 10 times before its capacity degenerates to an ineffective level. The condensation and treatment of the steam utilized in the regeneration step needs to be addressed as a separate operation. This regenerative technique has a better chance of success if there is a single contaminant in the water.

As noted above, one carbon replacement option is to have the spent material thermally reactivated off-site and returned to service. Thermal reactivation is conducted in a furnace where temperatures of up to 1800°F are obtained. The advantage of thermal reactivation is that the organic contaminants are driven off the carbon and thermally destroyed. The regeneration facility should be equipped with a thermal afterburner to insure complete destruction of organic compounds, and a scrubber to remove acid gases that may be present due to chlorinated solvents often found in groundwater.

The advantage of regeneration is the recovery of the granular activated carbon for further use. The regeneration step may be a custom operation, in which the spent carbon is regenerated as a batch and returned to the user. Regeneration may incur a 10 to 20% material loss, which must be made up to maintain the same amount of carbon on-line. Pool regeneration involves collection and regeneration of many small carbon units to effect economy of scale, but the carbon returned is not the original material.

Virgin carbon should be used in applications involving treatment of the water for reuse (potable applications for example) but reactivated carbon is generally less expensive in applications involving aquifer restoration where the water is treated for discharge.

Regenerated carbon may also possess different adsorptive qualities than does virgin grade material. These qualities may or may not be detrimental to the efficiency of contaminant removal. A proper evaluation of the regenerated material, followed by an overall economic comparison of the cost and effectiveness of the two processes, virgin carbon versus regenerated carbon, will determine the most economical approach.

Many groundwater applications require only small amounts of activated carbon. The "55 gallon" drum design is widely used in groundwater treatment systems. While the sizes vary, it normally requires a minimum of a 1000 to 2000 lbs of carbon in order for the carbon company to economically pick up and regenerate the spent carbon. Therefore, smaller units cannot take advantage of a regeneration service as part of the purchase of activated carbon.

There are three choices for carbon disposal when using the small units: (1) stockpile and regenerate; (2) landfill; and (3) incineration. The first choice is to accumulate enough carbon to send to a regeneration facility. This can be done by combining spent carbon from one or more remediations. The main problem with this scenario is that the spent carbon can be considered a hazardous waste, and could fall under regulations for hazardous waste storage. Even when the several locations combine to regenerate their carbon, each individual carbon canister will have to be analyzed before regeneration. The carbon regeneration companies need this information to ensure that there are no compounds present that will not be destroyed or captured by their regeneration system. The detailed analysis will add significant increases to the cost of regeneration.

The second method for disposal is to place the spent carbon in a landfill. Once again, analysis will have to be performed on each drum of spent carbon. In this case, the tests are needed to ensure that the spent carbon will meet all regulations concerning solid waste placed in a hazardous waste landfill. The land disposal restrictions may significantly limit options for landfilling spent carbon.

A second problem with placing spent carbon in a landfill is the nature of the adsorption process. As we discussed in the beginning of this section, carbon adsorption is an equilibrium process. Carbon isotherms are based upon the concentration in equilibrium between the carbon and the water. If the spent carbon comes into contact with clean water, equilibrium will be established, and some of the material will be exchanged to the water. Therefore, a landfill is not a good long-term solution for disposal of spent activated carbon.

The third method for disposal is incineration. In this case, both the carbon and the hazardous waste are destroyed. While this solution can be expensive, small quantities are no problem, and future liabilities are eliminated.

Operating Results—Case Studies

Granular activated carbon has been utilized successfully in many cases to treat contaminated groundwater. The case studies presented here show some of the wide variety of organic compounds and concentrations that may be present in groundwater that can be removed effectively and economically by granular carbon adsorption.

Table 3-11 summarizes actual cases in which carbon was used to treat a contaminated groundwater for drinking water use. All of these cases involved use of virgin-grade granular carbon to ensure the purity of the water.

This review shows that in cases where one contaminant is present in substantial quantity, effective use of the carbon can be obtained with short contact times (less than 30 minutes) and a single fixed bed system. This design provides for minimal equipment and carbon on-line, and subsequent minimum cost of treatment. In a vast majority of cases involving contaminant concentrations at μg/l levels, carbon usage rates are less than 0.5 lbs/1000 gallons.

If the carbon treatment is being utilized for drinking water purposes, chlorination is usually used after the adsorption process, to ensure no biological activity in the downstream distribution system. The operating costs for treating groundwater contaminated at μg/l levels range from 22 to 55 cents/1000 gallons. These operating costs reflect the amortization of the installed equipment and the replacement of the granular activated carbon as required.

Table 3-12 shows cases of groundwater contamination in which contaminants are present in mg/l levels. These examples include cases in which chemical spills, landfills, and storage tanks have led to severe groundwater problems. The situations covered here utilized reactivated grade carbon, as the end use of the treated groundwater is not for drinking water, but for discharge to surface water, recharge to the aquifer, or plant process use.

Due to the higher concentrations, and in most cases the presence of two or more predominant contaminants, the process of choice becomes the staged system, which ensures more complete utilization

TABLE 3-11 Carbon Adsorption with ppb Influent Levels[a]

System No.	Contaminants	Typical Influent Conc. (μg/liter)	Typical Effluent Conc. (μg/liter)	Surface Loading (gpm/ft²)	Total Contact Time (min)	Carbon Usage Rate (lb/1000 gal)	Operating Mode
1	1,1,1-Trichloroethane	143	<1	4.5	15	0.4	Single fixed bed
	Trichloroethylene	8.4	<1				
	Tetrachloothylene	26	<1				
2	Methyl T-butyl ether	30	<5	5.7	12	0.62	Two single fixed beds
	Di-isopropyl ether	35	<1				
3	Chloroform	400	<100	2.5	26	1.19	Four single fixed beds
	Trichloroethylene	10	<1				
4	Trichloroethylene	35	<1	3.3	21	0.21	Three single fixed beds
	Tetrachloroethylene	170	<1				
5	1,1,1-Trichloroethane	70	<1	4.5	30	0.45	Two fixed beds in series
	1,1-Dichloroethylene	10	<1				
6	Trichlorethylene	25	<1	2.0	35	0.32	Single fixed bed
	Cis-1,2-dichloroethylene	15	<1				
7	Trichlorethylene	50	<1	1.6	42	0.38	Two single fixed beds
8	Cis-1,2-dichloroethylene	5	<1	1.91	70	0.25	Two fixed beds in series
	Trichloroethylene	5	<1				
	Tetrachloroethylene	10	<1				

[a]From O'Brien, R. and Ficher, J. L. "There is an answer to groundwater contamination." *Water/Engineering & Management*, May 1983.

TABLE 3-12 Carbon Adsorption with ppm Influent Levels[a]

System No.	Contaminants	Typical Influent Conc. (mg/liter)	Typical Effluent Conc. (mg/liter)	Surface Loading (gpm/ft²)	Total Contact Time (min)	Carbon Usage Rate (lb/1000 gal)	Operating Mode
1	Phenol	63	< 1	1.0	201	5.8	Three fixed beds in series
	Orthochlorophenol	100	< 1				
2	Chloroform	3.4	< 1	0.5	262	11.6	Two fixed beds in series
	Carbon tetrachloride	135	< 1				
	Tetrachloroethylene	3	< 1				
	Tetrachloroethylene	70	< 1				
3	Chloroform	0.8	< 1	2.3	58	2.8	Two fixed beds in series
	Carbon tetrachloride	10.0	< 1				
	Tetrachloroethylene	15.0	< 1				
4	Benzene	0.4	< 1	1.21	112	1.9	Two fixed beds in series
	Tetrachloroethylene	4.5	< 1				
5	Chloroform	1.4	< 1	1.6	41	1.15	Two fixed beds in series
	Carbon tetrachloride	1.0	< 1				
6	Trichloroethylene	3.8	< 1	2.4	36	1.54	Two fixed beds in series
	Xylene	0.2–0.5	< 1				
	Isopropyl alcohol	0.2	< 10				
	Acetone	0.1	< 10				
7	Di-isopropyl methyl phosphonate	1.25	< 50	2.2	30	0.7	Single fixed bed
	Dichloropentadiene	0.45	< 10				

[a]From O'Brien, Robert, and Ficher, J. L. "There is an answer to groundwater contamination." *Water/Engineering & Management,* May 1983.

of the activated carbon while maintaining the effluent at the desired level. The multistage system also allows for the longer contact times which are required to meet the low effluent concentrations.

In some cases, pretreatment to the carbon system is necessary. Filtration may be required if the water is high in suspended solids material or iron. As we discussed during the air stripper section, iron in the groundwater can also precipitate in the activated carbon. The carbon bed should not be utilized as a filter. The carbon is designed as a column operation in the first place to get the maximum use of the carbon. If the bed is backwashed to removed suspended solids, then the carbon will be mixed and the transfer zone destroyed. If we lose the transfer zone, then the carbon will only perform as it did with the isotherm tests. For full utilization of the carbon, then, the bed must maintain its integrity.

pH adjustment may also be required if the water has a high pH and contains mineral salts susceptible to precipitation in the carbon bed. These pretreatment needs would normally be determined in the evaluation procedures.

The operating costs for treating groundwater contaminated at mg/l levels range from 38 cents to $2.52 per 1000 gallons. As before, these costs reflect amortization of installed equipment and replacement of the granular activated carbon as required. In most of these cases, the replacement of carbon will require regenerated material, which reflects some savings over utilizing virgin-grade material.

These results, showing removal of a wide range of organic contaminants to low or nondetectable levels, indicate that granular carbon adsorption is a versatile groundwater treatment process.

Application with Other Technologies

Carbon adsorption is a relatively expensive process. However, the inherent advantages of the technology make it particularly suited for low concentrations of nonvolatile components, high concentrations of nondegradable compounds, and short term projects. When there is a variety of compounds, or when very low effluent levels are required, carbon adsorption can be combined with other treatment techniques for the effective implementation of a groundwater remediation program.

Carbon adsorption may be readily combined with biological treatment to effect better overall performance. Powdered activated carbon may be added directly to the biological system, both provide sites for organic compounds to adsorb and undergo biological degradation, or to remove refractory organic compounds that may be toxic to the system. For possible toxic compounds, an evaluation procedure is conducted, considering specific organic compounds only. Finally, granular carbon systems can be used to polish the effluent from biological systems to remove refractory compounds.

Carbon adsorption also serves as a complimentary technology to air stripping. Granular activated carbon systems can be utilized to treat air stripping effluent water to remove remaining volatile and nonvolatile components. Nonvolatile components such as phenols, pesticides, and other substituted aromatics can be removed in a carbon adsorption step. As air stripping is an equilibrium process, there will be some concentrations of the volatile contaminant remaining, following the treatment step, so carbon can be utilized to remove such contaminants to nondetectable levels. The utilization of air stripping as pretreatment to carbon adsorption increases the life of the carbon. As the more volatile contaminants tend to be those less readily absorbed, their removal allows for use of less carbon. In many cases, the application of both air stripping and granular activated carbon will be the most cost-effective solution.

Although this section has discussed the treatment of contaminated waters, the application of granular activated carbon to treat gas streams is of importance as a complementary technology. As discussed in the Air Stripping section of this chapter, the use of air stripping may, in some cases, result in an unacceptable emission of organic compounds into the atmosphere. Granular activated carbon has been proven to be effective in removing organic vapors from such exhaust air streams.

Granular activated carbons and systems for vapor adsorption are different from those normally used for liquid phase systems. The carbon particles are usually larger to minimize pressure drop of the gas stream, and as contaminants are easily volatilized, the systems can be designed for in situ regeneration. The pore distribution of vapor phase carbon favors adsorption pores, therefore, equilibrium capacities for organic contaminants will be higher for vapor phase adsorption. The evaluation of activated carbon for vapor phase adsorption is similar, and isotherms for a variety of contaminants have been established. Pretreatment with condensers or dehumidifiers will also enhance the vapor phase adsorption step by reducing water vapor content and reducing the volume of the gas to be treated.

CHEMICAL OXIDATION

The use of chemical oxidation processes in the destruction or detoxi-fication of the contaminants found in groundwater has been practiced for hundreds of years. The use of chemical oxidation processes in the treatment of groundwater offers distinct advantages over other technologies. Recent improvements in chemical oxidation methods are increasing the application of this technology to the treatment of groundwater.

Oxidation processes involve the exchange of electrons between chemical species and effect a change in the oxidation (valence) state of the species involved. Specifically, oxidation processes are referred to as oxidation-reduction (redox) reactions because one of the species involved gains electrons (reduced valence state-reduction) and another loses electrons (increased valence state-oxidation). This exchange of electrons will destroy organic compounds by breaking carbon bonds and creating new, smaller compounds.

Three chemical oxidants have been widely used in industrial and groundwater treatment processes: chlorine, ozone, and hydrogen peroxide. In addition, oxygen has been used for some simple oxidation situations, i.e., iron removal.

Chemical oxidants have been used for the oxidation of organic and inorganic compound in the treatment of industrial wastewaters. Principal industrial uses for chemical oxidants include metals precipitation (iron, chromium) and liquid and gas treatment (destruction of cyanides, sulfides), and disinfection.

The principal reason for the use of chemical oxidation in the treatment of groundwater is the ability of oxidizing agents to degrade carbonaceous compounds, theoretically to carbon dioxide and water. Adequate oxidant must be present to facilitate a complete reaction.

Principles of Oxidation

Let us review the basic oxidation-reduction reaction. A redox reaction may be separated into the oxidation and reduction half-reactions, as presented here in the oxidation of ferrous iron using hydrogen peroxide (under acidic conditions):

$$H_2O_2 + 2H^+ + 2e^- \rightarrow 2H_2O \qquad (3\text{-}6)$$

$$2(Fe^{2+} + 3H_2O \rightarrow Fe(OH)_3 + e^- + 3H^+) \qquad (3\text{-}7)$$

In this example the addition of hydrogen peroxide to a solution which contains ferrous iron molecules causes an electron to be stripped from the iron atom. The peroxide molecule then combines with hydrogen atoms and assumes a more stable (lower energy) form as two water molecules.

The relative strength of an oxidant is commonly described by its electrode potential E°. Table 3-13 presents a summary of the standard electrode potentials for the oxidants generally employed in the treatment of groundwater. The values of E° for the oxidation and reduction half-cell elements of a redox reaction may be summed to calculate the thermodynamic potential of the reaction. This defines the level of energy input required or released during a reaction. The use of thermodynamic relationships for inorganic and some simple organic oxidations corresponds acceptably to theory, while complex organic redox reactions tend to be driven by chemical kinetics.

If a particular redox reaction occurs readily under the standard temperature and chemical setting of the groundwater, then the addi-

TABLE 3-13   Standard Electrode Potentials for Chemical Oxidants Used in Groundwater

Oxidant	Reduction Half-reaction	$E, ^\circ V$
Chlorine	$Cl_2(g) + 2e^- \rightarrow 2Cl^-$	1.36
Hypochlorous acid	$HOCL + H^+ + 2e^- \rightarrow Cl^- + H_2O$	1.49
Hypochlorite	$ClO^- + H_2O + 2e^- \rightarrow Cl^- + 2OH^-$	0.90
Ozone, acidic	$O_3 + 2H^+ + 2e^- \rightarrow O_2 + H_2O$	2.07
Ozone, basic	$O_3 + H_2O \rightarrow O_2 + 2OH^-$	1.24
Hydrogen peroxide		
Acidic	$H_2O_2 + 2H^+ + 2e^- \rightarrow 2H_2O$	1.78
Basic	$HO_{2^-} + 2e^- + H_2O \rightarrow 3HO^-$	0.85
Chlorine dioxide	$ClO_2 + 2H_2O + 5e^- \rightarrow Cl^- + 4OH^-$	1.71
Oxygen		
Acidic	$O_2 + 4H^+ + 4e^- \rightarrow 2H_2O$	1.23
Basic	$O_2 + 2H_2O^+ + 4e^- \rightarrow 4HO$	0.40

tion of the oxidant to the oxidate is all that is required. However, many redox reactions require the input of energy in the form of heat, ultraviolet (UV) light or chemical additions (such as pH alteration) or the presence of catalysts to economically facilitate a desired reaction.

Chemical Oxidants

As discussed previously, oxygen, chlorine, ozone, and hydrogen peroxide comprise the vast majority of oxidants used in potable water, industrial water, and wastewater treatment applications. These four oxidants are reviewed below.

Oxygen

Oxygen was the first oxidant used by man and it is the predominate oxidant used in nature for the destruction of wastes. In groundwater treatment, air (containing 21% oxygen) is principally used for the oxidation of ferrous iron to form insoluble ferric hydroxide. Atmospheric oxygen is also used in biological and other methods employed for groundwater treatment. The main cost of oxygen arises because of the need for power for transferring the oxygen from the atmosphere to the water.

Chlorine

Chlorine, a powerful and widely used oxidant in water and wastewater treatment industries, has seen limited application in the treatment of groundwater. This is because of the generation of chlorinated products and by-products. The proliferation of chlorinated species often renders the groundwater unsuitable for required purposes.

Chlorine is the most common oxidant used in water treatment. Chlorine is available in gaseous form in pressurized metal containers and is also available as a concentrated aqueous solution (sodium hypochlorite) or as a solid (calcium hypochlorite). Once added to water, the reaction chemistry of the various forms is essentially the same. The primary use for chlorination is to kill bacteria in potable water supplies. Chlorine maintains a lasting residual concentration in closed water systems, thus providing an extended disinfection ability. The main costs of chlorine arise from chemical, transportation, and storage needs.

Ozone

Ozone is the strongest of the oxidizing agents presented in Table 3-13. Ozone occurs naturally in the earth's atmosphere by the reaction of oxygen with UV radiation from the sun, and during thunderstorms with energy from lightning. Ozone is generated at the point of use with an apparatus which applies electric current to generate an electromagnetic field. The plasma or corona discharge excites oxygen molecules to the highly unstable ozone form. The specific yield of ozone generated by these devices is dependent on the applied voltage, frequency, design of the ozonator, and the type of feed gas used.

Ozone has properties which reduce its effectiveness as an oxidant for groundwater treatment. Ozone is so reactive that it will dissipate rapidly after contact with water, either by reacting with the impurities in the water or by spontaneous decomposition. Ozone decomposition is a complex chain-reaction process which occurs when ozone comes in contact with organic and inorganic molecules, then strips electrons, thus permitting the ozone to assume more stable forms such as elemental oxygen, hydroxide molecules, and water. One ozone decomposition intermediate is the hydroxyl radical (HO), one of the most powerful oxidizing agents known. The HO radical is capable of oxidizing almost any organic compound.

The capital and operating costs associated with the use of ozone as a chemical oxidant often limit the use of ozonation technologies. The small quantities of impurities customarily present in groundwater frequently require high levels of ozone addition for treatment, most of this ozone then decomposes spontaneously. The main cost of ozone is the capital and operating cost of the ozone generator.

Hydrogen Peroxide

Hydrogen peroxide is a stable and readily available substance and can effectively oxidize many compounds. Hydrogen peroxide is available in various commercial purities, ranging generally from 30 to 70% purity and in water-based solutions. Because of the relative stability of the material, it can be stored in metal, glass, and in some types of plastic containers.

Hydrogen peroxide is used commercially as an oxidant for numerous organic and inorganic materials, in both aqueous and vapor form. Hydrogen peroxide is not flammable, although its strength as an oxidizer encourages the combustion of flammable materials. Care is required to prevent contact or contamination of hydrogen peroxide with organic substances. Reactive decomposition is accelerated with exposure to high temperature or UV radiation.

The rates of redox reactions involving hydrogen peroxide are often increased with the use of catalysts. Sometimes reaction products or by-products serve as catalysts and these compounds are referred to as autocatalysts. Iron can serve as the catalysts for many hydrogen peroxide reactions, but catalyst additions should always evaluated for this method. The capital and operating costs associated with the use of hydrogen peroxide limit its application in ground-water treatment, although an increase in efficiency and effectiveness of hydrogen peroxide as an oxidant is observed when ultraviolet light is introduced.

Oxidation Reactions

As discussed, numerous factors affect the selection and application of oxidants to groundwater treatment problems. The principle limitations on the cost effectiveness arises because of the concentration of contaminants in the groundwater, and the quantity of excess oxidant needed to overcome spontaneous and unproductive decomposition in the aqueous medium. The reaction kinetics of a given redox reaction determine the degree of effectiveness available and the quantity of outside stimulation (in the form of a catalyst or energy input) needed to achieve an efficient and cost-effective solution.

Redox reaction rates which have been developed on the basis of the results of laboratory assessments are subject to numerous limitations because of variations in temperature, pH, the formation of reaction intermediates, and the presence and effectiveness of catalysts. Of particular concern in the recent application of chemical oxidation to groundwater treatment are the effect of temperature on reaction rates and the long-term effectiveness of the catalysts or energy sources used to stimulate reactions.

Ground-water temperature and the degree of temperature fluctuation characteristic of the source exhibit a large influence on the effectiveness of a redox process. As a rule of thumb, chemical reaction rates double with a 10°F increase in temperature. If a ground-water changes temperature on a seasonal basis, or experiences changes in temperature associated with aquifer depletion or ground-water age, effectiveness could increase or decrease. Besides the temperature of the groundwater the energy demands exerted by the introduction of the oxidant into the groundwater and also the effectiveness of oxidant mixing must be considered.

The effects of pH on reaction rate and effectiveness include changes in the redox reactions and on the quantity of available oxidant resulting from a given addition rate. Most chemical oxidants have different reaction mechanisms under acidic or alkaline conditions (Table 3-13) and have characteristic rates of reaction based on pH level. The selection of a chemical oxidation technology must consider the groundwater pH and also changes in geochemistry anticipated during the life cycle of the remediation.

Advanced Oxidation Processes

One of the most exciting oxidation technologies that is emerging as a viable treatment technique for groundwater is UV-Oxidation. This is a technology that should prove to destroy many organic compounds in a short amount of time and at a reasonable cost. EPA's Superfund Innovative Technology Evaluation (SITE) program has evaluated this technique and has published positive results about it. Several companies have created major marketing campaigns to promote UV-Oxidation to regulators, industries, and consultants in the groundwater field. The result has been that several UV-Oxidation projects are now proposed and some are being installed for treatment of ground water.

There are two basic forms in which UV-Oxidation are being applied: UV-Ozone and UV-Peroxide. Figure 3-27 shows a typical set up for a UV-Ozone System. Figure 3-28 shows a typical set up for a UV-Peroxide system. Both systems use an oxygen-based oxidant, ozone for the first and hydrogen peroxide for the second. UV light is used in conjunction with the oxidant. The UV light bulbs are placed in the reactor where the oxidant comes into contact with the contaminants in the ground water.

While ozone and hydrogen peroxide are both strong oxidizing agents, their effectiveness increases dramatically when stimulated by UV light. Figure 3-29 is an example of the difference between oxidation with the ozone alone and ozone stimulated with UV light. Similar types of increases are seen with UV and hydrogen peroxide. In both cases, the key to fast reaction time is the UV light source. However,

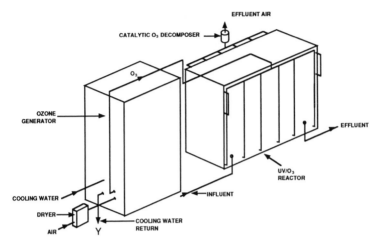

FIGURE 3-27. UV/ozone process flow diagram.

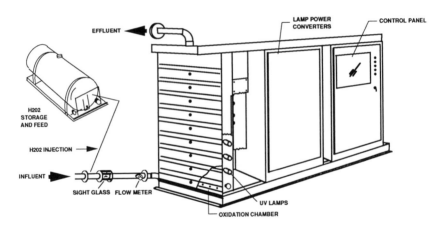

FIGURE 3-28. UV/H_2O_2 process flow diagram.

the source can not come into direct contact with the water. The bulbs are normally covered by a quartz tube. The quartz protects the bulbs, but allows the UV light to enter the water unaffected.

The main difference between the two designs is the type of oxidant and the method of application. Ozone is an unstable gas. It must be added to the reactor as small bubbles, and it must be produced at the site. The UV-ozone system includes an ozone generator. The ozone is

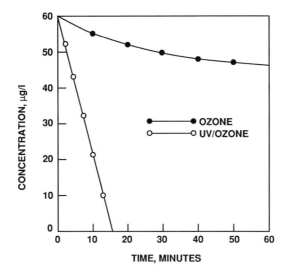

FIGURE 3-29. Oxidation of lindane with ozone and UV/ozone.

sparged into the reaction tank below the UV lights. This creates a gas stream that must be evaluated for ozone and volatile organic compounds. The design in Figure 3-27 and the unit that was studied under the SITE program use "low" intensity UV bulbs.

By comparison, hydrogen peroxide is a relatively stable liquid that can be delivered and stored on-site. The hydrogen peroxide is metered into the influent and the main reaction occurs within the reaction tank where the UV light is present. The design in Figure 3-28 uses "high" intensity UV bulbs. The reaction tank can be pressurized because the gas transfer requirements that apply to ozone gas do not apply in this case.

While these two technologies offer great promise for treatment of organic hazardous waste, there are potential problems with full-scale applications. The problems seem to be centered on the quartz tubes. Many chemicals and minerals in the groundwater coat out on the quartz tubes and prevent the UV light from getting into the water. This should not be unexpected with groundwater cleanups. Shallow aquifers are not normally used as sources of drinking water, because they commonly contain high concentrations of dissolved minerals and other dissolved suspended solids. Most groundwater contamination is found in shallow aquifers. There will be many natural chemicals in the water along with the contaminant that will interfere with any groundwater treatment process.

The laboratory methods that are being used to evaluate UV-Oxidation performance are limited. We must remember that the results only reflect what the optimum reaction rate can be. The main factor that needs to be included in the current examination protocol is the effect of the ground water on the quartz tubes. At the present time, the only way to do that is to run a long-term (1 to 3 months) pilot test on the actual ground water to be treated. Hopefully, someone will develop a short-term test for the quartz tubes in the near future.

MISCELLANEOUS TREATMENT METHODS

New technologies are constantly being introduced to the groundwater market. Most of these technologies were previously utilized in industrial processes. Some are completely new to the area. As these systems are applied in the field, more information will be published about each technology. The reader will have to be constantly updating the knowledge base.

In general, a shift in technological development has been evident in the last several years. Treatment systems are moving away from a phase change approach to a destructive treatment approach. Constituent removal and total destruction are more desirable because they not only remediate the process stream but they also eliminate the liability associated with phase change technologies. Some of the more interesting and significant new treatment methods are covered below.

Thermal Oxidation

Incineration is the raising of both water and contaminants to high temperatures in the presence of oxygen, resulting in carbon dioxide, water, and other products of combustion. One application of thermal oxidation is in the use of flame incinerators, where either a sufficient quantity of volatile organic compounds (VOCs) or supplemental fuel is used to maintain a flame burning in the 1500 to 2000°F range. When waters contain approximately 20% organics, the contaminated water will have a self sustaining burn. At concentrations below this level, auxiliary fuel is required. At lower concentrations this can become very expensive. Between adsorption in the unsaturated zone and dilution by the groundwater, concentrations are rarely this high in groundwater. These incinerators, while effective and in current use, require an intensive trial burn and permitting process. Another problem is the high capital cost of incinerators. Finally, groundwater remediation usually contains too small a volume of highly contaminated water. If the groundwater site is close enough to an existing incinerator, then the application for small volumes of highly toxic materials may be possible. There are also a few portable incinerators available.

A development in incineration technology is the nonflame, packed bed thermal incinerator. Field applications to date have included chlorinated hydrocarbons to a destruction efficiency of 99.99%. The corrosion resistant, continuous feed unit is packed with ceramic beads which are preheated to temperatures between 1900 to 2000°F. Contaminated gases, in the presence of oxygen, are forced through the heated ceramic beads and are mixed by turbulence allowing complete combustion to be achieved. Levels of nitrous oxides (NO_x) are reduced and the formation of furans and dioxins minimized by the uniform heating conditions below 3000°F. Additionally, safety devices required in open-flame units are not needed. A broad range of air flows may be accommodated with these units from 1 cubic foot per minute (cfm) to > 1000 cfm.

Although expensive, capital costs for nonflame, packed bed incinerators are comparable to on-site regenerative carbon systems. When compared to off-site carbon regeneration, costs are attractive. Care must be taken in evaluating applicability, as final costs will be influenced by fuel needs and influent stream complexities. Nonflame incinerators have not been successful in treating fluorinated compounds.[22] Once again, groundwater is usually too clean to make use of this technology.

One area that has been using incineration is the destruction of toxic chemicals in soil. This treatment is limited to cases where only the top layers of soil have been contaminated. Also, the contaminants are not soluble in water and therefore have remained at the surface, where they were originally released, and therefore a limited amount of contaminated soil has been created. Soil incineration is an expensive process, and unique situations must exist in order for it to be used. On-site incineration can be used when the engineer decides that the contaminated soil is too dangerous to transport over the roads, or when the nearest landfill is too far away.

Probably the most widely applied thermal method is catalyzed thermal oxidation, using either metals or inorganic acid as the catalyst. These units are applied to air treatment from vapor extraction systems or off-gas treatment from air strippers. Metal catalyzed oxidation is designed for destruction efficiencies from 90 to 98% and is a function of the temperature of the catalyst bed, quantity of catalyst, and type of metal used.[23] The contaminated air stream is preheated with reuse of waste heat and a burner, if required. Preheated air passes through the metal catalyst which, in the presence of oxygen, promotes combustion reactions with temperatures ranging between 500 and 1000°F. The catalyst surface lowers the activation energy required to cause the oxygen to react with the contaminant hydrocarbons.[24] Products of combustion, water and carbon dioxide, are emitted by stack discharge after passing through a heat exchanger, as indicated earlier.

The difference between metal catalyst lies in the component material and the physical structure. Base metals such as copper (Cu), manganese (Mn) and cobalt (Co) are used, as are precious metals including platinum (Pt), palladium (Pd) and rhodium (Rh). Selection of less expensive base metals would be appropriate for fluidized bed reactors which require periodic replenishment of the metal catalyst. Although expensive, precious metal catalysts have been found effective at lowering operating temperatures and additionally are resistant to contaminants.

The physical forms of metal catalysts which have been successfully used are in pelleted and honeycomb forms. Pelleted forms, used in fluidized beds, must operate within a narrow contaminant gas flow rate to eliminate channeling and bed collapse, and therefore require carefully controlled, operator assisted start-up. Although the catalyst is not consumed in the combustion reaction, breakdown of the pellets occurs in a fluidized bed because of friction. The last catalyst has to be continually replaced.

Catalyst performance is influenced by a number of conditions: thermal aging, poisoning, and masking of the catalyst. Thermal aging reduces the catalyst activity by sintering of the surface area in base metals because of site closure. In precious metals, sintering of the surface area occurs because of crystallite migration. Sintering occurs with temperatures in excess of 1400°F, an unusual condition in catalytic thermal oxidation.

Catalyst poisoning occurs when contaminants in the waste stream react with catalyst metal sites, and thus reduce catalytic activity. Potential poisons include halogens and phosphorous. For precious metals, poisoning is an adsorptive phenomenon which may be reversed by desorption, while base metals poisoning is irreversible because of the inability of the catalyst to be cleaned with chemical solutions which solubilize the base metals.

Masking of the metal catalyst is the covering or masking of available catalytic sites, this reduces catalyst activity. This condition occurs when the process is operated at too low a temperature, thus resulting in char formation. Masking compounds may be removed with either acid or base solutions or a combination of both.

Low temperature catalytic oxidation is selected when influent stream organic concentrations are low, requiring supplemental fuel for flame thermal incineration. A benefit to also be considered in catalytic thermal oxidation is lowered nitrous oxide (NO_x) and carbon dioxide emissions associated with lower fuel consumption.

Chemical Oxidation

The previous section in this chapter was devoted to oxidation methods. One additional new oxidation method, liquid phase oxidation using an inorganic acid catalyst is an emerging technology. In this method, the contaminated liquid stream is mixed with an inorganic acid solution such as phosphoric acid, an oxidant, and heated to temperatures in the range of 250 to 500°F. The acid catalyzed oxidation reaction results in dissolution of contaminants, oxidant, and catalyst in the liquid phase. The by-products of this reaction mechanism are nitrogen, oxygen, carbon monoxide, carbon dioxide, hydrogen, and water, with the ratio of carbon monoxide to carbon dioxide in the 33% range. Addition of a transition metal in small quantities has been shown to improve catalytic action and achieve destruction efficiencies greater than 99%. This technology has been proven effective in the reduction of pesticides and a herbicide.[25]

Membrane Technology

Two types of membrane technology are noteworthy: diffusion membrane separation and pervaporation membrane separation. Diffusion membrane separation involves a nonporous membrane, which unlike a reverse osmosis membrane is permeability selective. The system operates with a low pressure gradient, approximately 35 pounds per square inch (psi) which causes diffusion of water through the membrane. Impermeability of hydrocarbons causes these compounds to be retained by the membrane. The structure and orientation of the membrane polymer chains affect the degree of permeability afforded by the membrane. Pilot studies have shown effectiveness on halogenated and nonhalogenated compounds, with less fouling than in standard filtration systems. Other advantages to this technology are the low operating pressure and the effectiveness in treating oily substances.

Like diffusion membrane separation, pervaporation membrane separation is a perm-selective technology. Liquid feed containing contaminants, contacts the membrane on one side of the membrane, and is removed as a vapor on the other. The phase change in commercial operations is accomplished by maintaining a vapor pressure on the permeate side which is less than that of the pressure of the liquid feed. The vapor pressure is created by the maintenance of a low pressure on the permeate side by spontaneous condensation of permeate vapor. Control of the liquid condensate conditions determines the vapor pressure. Solvent removal of benzene in bench scale experiments indicates a 99% efficiency. The effluents of this system are a purified groundwater as well as a condensed, concentrated permeate liquid. The necessity of disposal and the associated liability of this phase transport technology is one of the main disadvantages of the technology.[26] Pervaporation is applicable to groundwater, leachate, or wastewater treatment. In comparison to carbon adsorption treatment, pervaporation does not entail competition between compounds in a multicomponent stream, for active adsorption sites. Compounds absorbed on the pervaporation perm-specific membrane are continuously removed as vapor. There are advantages for pervaporation when compared to air stripping. As no air is injected into the pervaporation system, the fouling problems experienced because of oxygen saturation do not occur. Additionally, the need for expensive air phase off-gas treatment is not required. One final advantage is that pervaporation is not limited to those compounds with high Henry's law constants. The major technical disadvantage to pervaporation treatment arises because of the need for the management of the concentrated permeate stream.[27]

Supercritical Extraction

Supercritical extraction utilizes a liquid or gas at or near its critical point to act as an enhanced solvent for removal of hazardous compounds from process streams. There is also evidence that supercritical liquid extraction is effective in the removal of aromatic hydrocarbons from sandy loam soil.

It is known that liquids or gases at or near their supercritical point have an increased solvent action while maintaining their original diffusivities and viscosities. Controlling the pressure and temperature will control the extraction and separation process by affecting the ability of the fluid to act as a solvent. The system influent, composed of a solid or liquid contaminant containing stream, is introduced into a reactor vessel and mixed with the supercritical solvent. After extraction of contaminants, the supercritical solvent/contaminant stream is drawn off to a separator vessel, where pressure reduction causes vaporization of the supercritical solvent and condensation of the contaminants. The contaminant stream is collected for disposal while the solvent is repressurized to its critical state for reuse.

Liquid carbon dioxide has been shown to be effective in the extraction of nonvolatile type compounds such as PCBs, pesticides, and phenol from both liquid and soil systems. Several factors and conditions increase the efficiency of extraction, including the addition of co-solvents to increase the solubility of carbon dioxide with multiring polyaromatic hydrocarbons (PAHs). Water has been shown to decrease the adsorptive ability of PAHs on soil, therefore improving its extractability under supercritical conditions. Generally, the more rings a PAH has, the greater difficulty there will be to achieve good extraction. Co-solvent addition has been shown to increase efficiency with multiring PAHs by 20 to 30%. Although supercritical extraction occurs at high pressures in the range of 900 to 1200 psi, moderate to ambient temperatures are possible for many applications.[28]

Emerging Techologies

A number of processes will now be briefly mentioned to complete this discussion of technologies. An attractive treatment for low concentration (μg/l) organics such as chloroform, TCE, PCE, substituted benzene, and trihalomethanes is high speed electron beam (E-beam) technology. E-beam technology has shown removal efficiencies up to 99.99% in full-scale operation. An electron beam, passing through a thin sheet of water initiates chemical reactions to reduce contaminants to carbon dioxide, water, and salt. Currently, this process has been accepted into the SITE program.[29]

Micellar enhanced ultrafiltration differs from conventional ultrafiltration by the addition of a surfactant to the waste stream. Contaminants are collected by the surfactant based micelles which are created through ionic attraction, and based on solubility are incorporated in the macromolecular structure. The ability to treat organics and/or heavy metals is influenced by the type of surfactant selected with respect to its ionic charge. Anionic, negatively charged surfactants, have the capability to absorb heavy metals, while positively charged cations are effective in organic removal but cannot remove metals. Removal efficiencies are reported at greater than 99% for divalent zinc, divalent copper, hexane, chlorophenol, and 4-tert-butylphenol, and greater than 97% for cresol. Solubilized contaminants and their micelles are collected in a concentrate stream for disposal, and the resulting clean filtrate is then available for use as desired. Some leakage in ultrafiltration of pure surfactant has been observed, but it is nontoxic and reports in the literature indicate that the concentration of the biodegradable substances is less than 100 ppm.[30]

References

1. Olsen, Roger and Davis, Andy, Predicting the fate and transport of organic compounds in groundwater. *Hazardous Materials Control,* May/June 1990.
2. Davis, J. B. et. al. *The Migration of Petroleum Products in Soil and Groundwater.* American Petroleum Institute, December 1972.
3. Yaniga, Paul M. Groundwater abatement techniques for removal of refined hydrocarbons. *Hazardous Wastes and Environmental Emergencies Proceedings,* March 1984. HMCRI.
4. Suchomel, Karen et. al. Production and transport of carbon dioxide in a contaminated vadose zone: A stable and radioactive carbon isotope study. *Environmental Science and Technology,* Dec. 1990.
5. Lenzo, F. C., Freilinghaus, T. J. and Zienkiewicz, A. W., The application of the Onda correlation to packed column air stripper design: Theory versus reality. *AWWA National Conference Proceedings.* 1990.

6. Sullivan, K. and Lenzo, F. 1989. Ground water treatment techniques—An overview of the state-of-the-art in America. *First US/USSR Conference on Hydrogeology,* Moscow (NWWA). July 3-5, 1989.

7. USEPA. July 8, 1987. US Federal Register, 40 CFR, Parts 141 and 142. National Primary Drinking Water Regulations. Pages 25690-25717.

8. Nicholson, B. C. et. al. Henry's law constants for the trihalomethanes: Effects of water composition and temperature. *Environmental Science and Technology,* 18:7-518, 1984.

9. Roberts, P. V., Hopkins, G. D., Munz, C., and Riojas, A. H. 1985. Evaluating two-resistance models for air stripping of volatile organic contaminants in a countercurrent, packed column. *Environmental Science And Technology.* 19:164-173.

10. Cummins, M. D. 1985. Economic Evaluation of TCE Removal From Contaminated Ground Water by Packed Column Air Stripping (DRAFT).

11. Cummings, M. D. 1988. Field Evaluation of Packed Column Air Stripping, Miami, FL. USEPA-ODW-TSD.

12. Cummings, M. D. 1984. Field Evaluation of Packed Column Air Stripping for THM Removal, Virginia Beach, VA. USEPA-ODW-TSD.

13. Staudinger, J., Knoche, W. R., and Randall, C. W., 1990. Evaluating the Onda mass transfer correlation for the design of packed column air stripping. *AWWA Journal,* pp. 73-79.

14. Carlton, G. M. 1986. Utilizing air stripping technology for pretreatment of solvent waste. *Proceedings of Solvent Waste Reduction Symposium.* Pages 14-32.

15. USEPA, Air Toxics Control Technology Center. 1987. Air Strippers Air Emissions and Controls (DRAFT). DCN No. 87-231-02032-16.

16. Crittenden, J. et. al. June 1987. An evaluation of the technical feasibility of air stripping solvent recovery process. American Water Works Association Research Foundation.

17. Zanden, A. K., Semmens, M. J., and Marbaitz, R. H. Removing VOC's by membrane stripping. *AWWA Journal.* November 1989. Pages 76-81.

18. Zanden, A. K., Qin, R., and Semmens, M. J. Membrane/oil stripping of VOC's from water in hollow fiber contactor. *ASCE Journal of Environmental Engineering.* 115(4), August 1989.

19. Foster, Robert; Lewis, Norma; Topudurti, Irankumar; and Weishans, Gary. A field demonstration of the UV/oxidation technology to treat ground water contaminated with VOCs. *Control Technology.* Air & Waste Management Assoc.

20. Bernardin, Frederick E., Jr. UV/peroxidation destroys organics in groundwater. 83rd Annual Meeting of the Air and Waste Management Association, Pittsburgh, PA. June 24-29, 1990.

21. Fletcher, David B., UV/ozone process treats toxics. *Waterworld News,* 3(3), May/June 1987.
22. Roy, Kimberly A., Scientists set to destroy VOCs with thermal oxidation process. *Hazmat World,* December 1989.
23. Herbert, Keith J., Catalysts for volatile organic compound control in the 1990's. Paper Presented at the 1990 Incineration Conference, May 1990.
24. Burns, Kenneth R., Use of catalysts for VOC control. Paper Presented at the New England Environmental Expo, Boston, Massachusetts, April 1990.
25. Leavitt, David et. al., Homogeneously catalyzed oxidation for the destruction of aqueous organic wastes. *Environmental Progress,* Vol. 9, November 1990.
26. Wijmans, J. G. et. al., Treatment of organic-contaminated wastewater streams by pervaporation. *Environmental Progress,* Vol. 9, November 1990.
27. Lipski, Chris and Cote, Pierre, The use of pervaporation for the removal of organic contaminants from water. *Environmental Progress,* Vol. 9, November 1990.
28. Andrews, Arthur T. et. al., Supercritical fluid extraction of aromatic contaminants from a sandy loam soil. *Environmental Progress,* Vol. 9, November 1990.
29. Roy, Kimberly A., High speed electrons race toward water cleanup. *Hazmat World,* December 1990.
30. A guide to innovative nonthermal hazardous waste treatment processes. Special Feature Article, *The Hazardous Waste Consultant,* November-December 1990.

4

Treatment of Organic
Contaminants
Biological Treatment

One of the most promising treatment technologies for groundwater is biological treatment. During the ten years that I have worked on groundwater, I have seen biotechnology go from snake oil to an advanced technology. Our government is now strongly behind bioremediation—"If we have a magical torch, it's biotechnology research," Reilly.[1]

With all of this attention, it is important to understand what biological treatment can really do. Biological treatment cannot be applied to every situation. We have to understand biotechnology's abilities and limitations before we broadly apply it to groundwater situations.

In order to correctly apply biotechnology, we have to understand the biochemical reactions of the microorganisms which are used, and we have to understand the equipment designs used to apply those microorganisms to groundwater. There is a large difference between what a bacteria can do with specific organic compounds and what an activated sludge treatment system can do with a specific groundwater situation. Accordingly, this chapter will start with a detailed review of microorganisms and their biochemical reactions with hazardous organic compounds. Next, we will review how these reactions have been applied to groundwater cleanups. As presented in Chapters 1 and 2, groundwater treatment will require special design considerations. Finally, we will discuss how these biochemical reactions can be applied directly in the ground or aquifer, or in situ treatment.

MICROORGANISMS

Free-living microorganisms that exist on earth include bacteria, fungi, algae, protozoa, and metazoa. Viruses are also prevalent in the environment, however, these particles can only exist as parasites in living cells of other organisms and will not be discussed in this book. Microorganisms have a variety of characteristics that allow their survival and distribution throughout the environment. They can be divided into two main groups. The eucaryotic cell is the unit of structure that exists in plants, metazoan animals, fungi, algae, and protozoa. The less complex procaryotic cells include the bacteria and cyanobacteria.

Even though the protozoa and metazoa are important organisms that affect soil and water biology and chemistry, they do not perform important degradative roles. Therefore, this chapter will concentrate on bacteria and fungi.

The bacteria are by far the most prevalent and diverse organisms on earth. There are over 200 genera in the bacterial kingdom.[2] These organisms lack nuclear membranes and do not contain internal compartmentalization by unit membrane systems. Bacteria range in size from approximately 0.5 micron to seldom greater than 5 microns in diameter. The cellular shape can be spherical, rod-shaped filamentous, spiral, or helical. Reproduction is by binary fission. However, genetic material can also be exchanged between bacteria.

The fungi which include molds, mildew, rusts, smuts, yeasts, mushrooms, and puffballs constitute a diverse group of organisms living in fresh water and marine water but predominantly in soil or on dead plant material. Fungi are responsible for mineralizing organic carbon and decomposing woody material (cellulose and lignin). Reproduction occurs by sexual and asexual spores or by budding (yeasts).

**Distribution and Occurrence of
Microorganisms in the Environment**

Microorganisms are found throughout the environment. Habitats that are suitable for higher plants and animals to survive will permit microorganisms to flourish. Even habitats that are adverse to higher life forms, can support a diverse microorganism population. Soil, groundwater, surface water, and air can support or transport microorganisms. Soil, groundwater, and surface water environments support microorganism growth while the air acts as a medium to distribute organisms to other environments. For example, there are several genera of bacteria and fungi in soil and water capable of hydrocarbon

TABLE 4-1   **Microorganism Population Distribution in Soil and Ground Water**

Organism	Population Size	
	Typical	Extreme
	Surface soil (cells/gram soil)	
Bacteria	0.1-1 billion	>10 billion
Actinomycetes	10-100 million	100 million
Fungi	0.1-1 million	20 million
Algae	10,000-100,000	3 million
	Subsoil (cells/gram soil)	
Bacteria	1,000-10,000,000	200 million
	Ground water (cell/ml)	
Bacteria	100-200,000	1 million

degradation. The predominant bacteria genera[5,6] include *Pseudomonas, Bacillus, Arthrobacter, Alcaligenes, Corynebacterium, Flavorbacterium, Achromobacter, Micrococcus, Nocardia,* and *Mycobacterium.* The predominant fungal genera[7] include *Trichoderma, Penicillium, Asperigillus, Mortierella,* and *Phanerochaete.* Table 4-1[13] shows the microorganism population distribution in soil and groundwater and demonstrates the variability and population sizes that can exist.

Soil

Bacteria outnumber the other organisms found in a typical soil. These organisms rapidly reproduce and constitute the majority of biomass in soil. Typically, microorganisms decrease with depth in the soil profile, as does organic matter. The population density does not continue to decrease to extinction with increasing depth, nor does it necessarily reach a constant declining density. Fluctuations in density commonly occur at lower horizons. In alluvial soils, populations fluctuate with textural changes; organisms are more numerous in silt or silty clay than in intervening sandy or coarse sandy horizons. In soil profiles above a perched water table, organisms are more numerous in the zone immediately above the water table than in higher zones.[3] Most fungal species prefer the upper soil profile.

Groundwater

Microbial life occurs in aquifers. Bacteria exist in shallow to deep subsurface regions but the origins of these organisms are unknown.

They could have been deposited with sediments millions of years ago, or they may have migrated recently into the formations from surface soil. Bacteria tend not to travel long distances in fine soils but can travel long distances in coarse or fractured formations. These formations are susceptible to contamination by surface water and may carry pathogenic organisms into aquifer systems from sewage discharge, landfill leachate, and polluted water.[4]

In all of these ecosystems, multiple types of bacteria exist. In fact, all degradation processes require multiple microorganisms working in concert. Also, more than one type of bacteria or fungi can perform the same degradation function. When investigating microorganisms for degradation of specific organic compounds, it is more important to demonstrate the ability to degrade a compound than to locate a specific bacteria or fungi.

Microorganism Biochemical Reactions

Microorganisms degrade organic compounds to obtain energy that is conserved in the C-C bonds of the compounds. The organics are converted to simpler organic compounds, and ultimately into carbon dioxide or methane, and water. The microbes will also use part of the compounds as building blocks for new microbial cells.

Microorganisms have been used by man to degrade organic compounds for many years. Biodegradation occurs when indigenous microorganisms convert or degrade natural and man-made organic compounds. Carbon sources not produced by any natural enzymatic process or having unnatural structural features are considered xenobiotic. Compounds that are naturally occurring and exist in increased concentrations as a result of human activities are also considered xenobiotic.[9]

The main goal of a bioremediation design is the destruction of organic hazardous waste. However, we must remember that this is not the main goal of the bacteria and the fungi. The main function of bacteria and fungi is the degradation of natural organic material. This, in turn, is part of the natural carbon cycle of the earth, Figure 4-1.

As can be seen in Figure 4-1, the microorganisms perform a small part of the overall carbon cycle. When we discuss hazardous waste destruction, we are referring to a small part of the natural microorganism activity. We are simply recycling man made carbon compounds back into the natural carbon cycle.

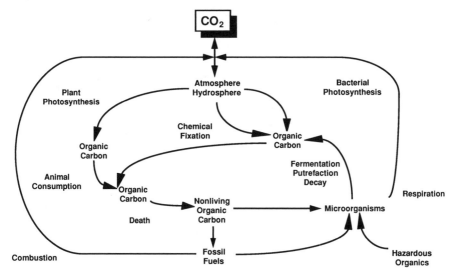

FIGURE 4-1. Carbon cycle.

It is helpful to keep these ideas in mind when we are designing biological systems. We are adjusting natural systems, not creating new ones. In fact, most of the time we will find that degradation is already occurring, and that the main design objective is to enhance an ongoing reaction.

Biodegradation of organic compounds (and maintenance of life sustaining processes) are reliant upon enzymes. There are numerous enzyme systems in bacteria which perform highly specific reactions. These biological reactions hasten and regulate cellular activity such as energetics and biosynthesis.

Enzymes that catalyze cellular reactions are proteins produced by living organisms. These proteins exist within cellular cytoplasm, or they may be attached to cellular membranes, or to the outside of the cell wall. Reactions are catalyzed when the organic substrate collides and binds to the active site of the enzyme. Substrate activation allows the enzyme to react and produce the product and restore the enzyme.[8]

The best way to understand enzyme reactions is to think of them as a lock and key. Figure 4-2 shows how only an enzyme with the right shape (and chemistry) can function as a key for the organic reactions. The key and lock in Figure 4-2 are two dimensional. In actuality the

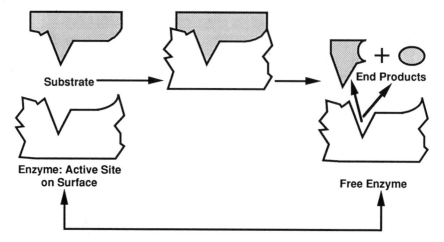

FIGURE 4-2. Enzymes are represented as a lock and key.

enzyme and organic chemical are three dimensional. The fit between the two has to be precise.

In the environment organic compounds that are degradable align precisely with the active site of the specific enzyme. Persistent compounds do not align well, and recalcitrant compounds do not bind with the enzyme's active site. Degradation of persistent or recalcitrant compounds requires that the microorganism population adapts in response to the environment by synthesizing enzymes capable of catalyzing degradation of those compounds.

The action of enzymes is limited by two basic factors. The physical constraints of the active site within the enzyme molecule result in a series of kinetic limitations referred to as enzyme inhibition. Extracellular enzymes are also susceptible to proteases (protein cleaving enzymes) that may be present in the environment.[9]

Organisms generally derive energy from oxidation-reduction reactions (catabolism). Enzyme mediated oxidation-reduction reactions are the transfer of electrons from electron donors to acceptors. Energy is derived from these reactions where the energy source (electron donor) is oxidized, transferring electrons to an acceptor and releasing energy conserved in the chemical bond. Once the electron donor has been completely oxidized, the compound is no longer a source of energy. Bioremediation processes where microorganisms are exploited to degrade xenobiotic compounds are *identical* to natural degradative processes requiring enzymes. The energy released from these compounds is used by the organism to maintain life sustaining processes.

Organisms conserve the energy generated during oxidation-reduction reactions by transferring the energy to high-energy phosphate bonds. The most important high energy phosphate compound in living organisms is adenosine triphosphate (ATP). ATP is generated and used during biosynthesis reactions which require energy expenditures.

ATP is synthesized in a variety of mechanisms. Organic and inorganic compounds, and light energy can be used to synthesize ATP. Organic compounds provide energy sources for all animals and most microorganisms including all fungi, protozoa, and most bacteria. The biochemical pathways in which ATP is generated can be divided into three major groups: (1) fermentation, in which oxidation occurs in the absence of added electron acceptors, (2) aerobic respiration, in which molecular oxygen serves as the electron acceptor, and (3) anaerobic respiration in which a compound other than oxygen, such as nitrate, sulfate, or carbonate, serves as the electron acceptor.[8]

Biosynthetic processes where microorganisms synthesize compounds that they require are termed anabolism. These are complex enzymatic processes in which simple compounds are synthesized into complex compounds. These processes require energy generated in the catabolic reactions. The combined catabolic and anabolic processes is called metabolism and refers to all degradative and biosynthetic reactions within cells.

Factors That Affect Biochemical Reactions

Several factors are necessary to maintain a microorganism's metabolic processes. Optimization of these factors will provide conditions that are conducive to support biodegradation. However, optimization is effected by factors that may be limited in the environment. Any factor (carbon, oxygen, inorganic nutrients) can limit the biodegradation rate of xenobiotic compounds. It is necessary, therefore, to establish an environment in which the limiting factor(s) are the xenobiotic compounds. This will assure that biochemical processes will be able to effectively degrade these compounds. The main objective of most above ground and in situ designs is to create an environment that does not limit the microorganism's rate of growth.

This process is illustrated in Leibig's law of the minimum, which states that the rate of biological processes is limited by the factor which is present at its minimum level. This law can also be extended to demonstrate that any growth factor can be toxic to biological processes if the concentrations are too high. Therefore, optimal conditions are such that necessary factors are not consumed or concentrations too high to inhibit growth. Even oxygen or water at too high a concentration can be harmful (toxic) to the biochemistry of the cell. On the other hand, the most toxic compound known will have a minimum concentration at which it no longer affects the microorganism. Figure 4-3 illustrates Leibig's law.

Electron Acceptor
As mentioned above, there are three mechanisms used by microorganisms to produce energy. Respiration processes require oxygen, fermentative processes rely on organic compounds as electron donors and acceptors, and anaerobic processes rely on nitrate, sulfate, or carbonate in the absence of oxygen to complete organic compound oxidation. Microorganisms that require molecular oxygen are termed obligately aerobic. These organisms cannot survive without oxygen. Within this group, there are microorganisms that survive on reduced oxygen concentrations and are termed microaerophilic. Microaerophilic organism's enzyme systems are actually inactivated under strong oxidizing conditions (partial pressure greater than 0.2 atm).[10] Micro-

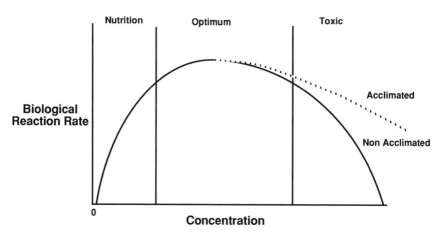

FIGURE 4-3. Leibig's law.

organisms that do not obtain energy using molecular oxygen and are inhibited or killed by molecular oxygen are termed obligately anaerobic. Microorganisms that can survive and grow either in the presence or absence of molecular oxygen are termed facultative anaerobes. In metabolic terms, facultative anaerobes are made of two subgroups. Some (lactic acid bacteria) have exclusively fermentative energy-yielding metabolism, but are not sensitive to the presence of oxygen. Others can shift from aerobic to anaerobic with the absence of molecular oxygen and in the presence of nitrates or sulfates.[10]

Generally, an oxygen atmosphere of less than 1% in soil will change metabolism from aerobic to anaerobic.[3] In aqueous environments, oxygen concentration less than approximately 1.0 mg/l can switch metabolism from aerobic to anaerobic.[11] Microaerophobic bacteria maintain aerobic reactions at reduced oxygen levels.

Inorganic Nutrients

The molecular composition of cells is fairly constant and indicates the requirements for growth. Water constitutes 80 to 90% of cellular weight and is always a major nutrient. The solid portion of the cell is made of carbon, oxygen, nitrogen, hydrogen, phosphorus, sulfur, and micronutrients. The approximate elementary composition is shown in Table 4-2.[10]

As can be seen from Table 4-2, the largest component of the bacteria is carbon. The organics that we wish to destroy will provide

TABLE 4-2   **Molecular Composition of a
Bacterial Cell**

Element	Percentage of Dry Weight
Carbon	50
Oxygen	20
Nitrogen	14
Hydrogen	8
Phosphorus	3
Sulfur	1
Potassium	1
Sodium	1
Calcium	0.5
Magnesium	0.5
Chlorine	0.5
Iron	0.2
Others	~0.3

this element. After carbon, oxygen is the highest percentage of the cell. When this is added to the oxygen which is required as the electron acceptor, large amounts of oxygen are utilized in biological degradation. In fact, oxygen is usually the limiting factor in biochemical reactions. Most of the operating costs of a biological system come from the need to supply oxygen to the bacteria.

The other major nutrients required by the microorganisms are nitrogen and phosphorous. This can also be seen in Table 4-2. Nitrogen is found in microorganisms in proteins, microbial cell wall components, and nucleic acids. The most common sources of inorganic nitrogen are ammonia and nitrates. Ammonia can be directly assimilated into amino acid synthesis. When nitrates are used, they are first reduced to ammonia and are then synthesized into organic nitrogen forms.

Phosphorous, in the form of inorganic phosphates, is used by microorganisms to synthesize phospholipids and nucleic acids. Phosphorous is also essential for the energy transfer reactions of ATP. Organic phosphate compounds occur in nature and are also used by microorganisms. Phosphatase enzymes that hydrolyze the organic phosphate ester are present in nearly all organisms.[8]

Micronutrients are also required for microbial growth. There are several micronutrients, such as sulfur, potassium, magnesium, calcium, and sodium, that are universally required. Sulfur is used to synthesize two amino acids, cysteine and methionine. Inorganic sulfate is also used to synthesize sulfur containing vitamins (thiamin, biotin, and lipoic acid).[8] Several enzymes including those involved in protein synthesis are activated by potassium. Magnesium is required for the activity of many enzymes, especially phosphate transfer and functions to stabilize ribosomes, cell membranes, and nucleic acids. Calcium acts to stabilize bacterial spores against heat and may also be involved in cell wall stability. Sodium is required by some, but not all microorganisms; bacteria that need sodium only grow in salt water environments.[8]

Other micronutrients commonly required by microorganisms include iron, zinc, copper, cobalt, manganese, and molybdenum. These metals function in enzymes and coenzymes. These metals (with the exception of iron) are also considered heavy metals and are toxic to microorganisms. Leibig's law is the best way to understand how these compounds can be both nutrients and possibly toxic.

All of these factors are necessary to maintain a microorganism's metabolic processes. One part of any biological design is the means to provide the nutrients that are required by the microorganism. Once again, the carbon source should be left as the limiting factor in the biochemical reaction.

Environmental Factors

There are several environmental factors that can also affect biochemical reactions. These factors can control the type of bacteria that are prominent in the degradation of organics and will affect the rate of degradation. The main environmental effects are temperature, water, and pH. We will also review the different factors that can lead to toxic or inhibiting conditions.

Temperature

Temperature is an important microorganism growth factor. As the temperature rises, chemical and enzymatic reaction rates in the cell increase. However, proteins, nucleic acids, and cellular components will become inactivated if the temperature becomes too high. For every organism there is a minimum temperature below which growth no longer occurs, an optimum temperature at which growth is most rapid, and a maximum temperature above which growth is not possible. The optimum temperature is always nearer the maximum than the minimum temperature. Temperature ranges for microorganisms are very wide. Some microorganisms have an optimum temperature as low as 5° to 10°C and others as high as 75° to 80°C. The temperature range in which growth occurs ranges from below freezing up to the boiling temperature of water. No single microorganism will grow over this entire range. Bacteria are frequently divided into three broad groups as follows: (1) thermophiles, which grow at temperatures above 55°C, (2) mesophiles, which grow in the midrange temperature of 20° to 45°C, and (3) psychrophiles, which grow well at 0°C. In general, the growth range is approximately 30 to 40 degrees for each group. Microorganisms that grow in terrestrial and aquatic environments grow in a range from 20° to 45°C. Figure 4-4 demonstrates the relative rates of reactions at various temperatures. As can be seen in Figure 4-4, microorganisms can grow in a wide range of temperatures. When designing a biological cleanup, the particular temperature is not as important as insuring that large temperature swings do not occur

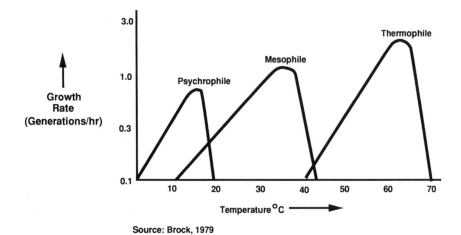

Source: Brock, 1979

FIGURE 4-4. Relationships of temperature to growth rate of a psychrophile, a mesophile and a thermophile.

during the project. Changes in temperature will cause a change in microbial population, and the new population will have to go through growth from the small numbers present when the temperature became optimum for it.

Water

Water may be the most important factor that influences microorganism growth. Water quantity and quality varies in different environments. Water functions to transport nutrients to the cells, aids the catalysts of many enzymes, and maintains turgidity of the cell (osmotic pressure). The availability of water to microorganisms can be expressed in terms of water activity, which is related to the vapor pressure of water in the air over a solution (relative humidity). Water activity in freshwater and marine environments is relatively high, and lowers with increasing concentrations of dissolved solute.[8] Bacteria can grow well in the salt water of an ocean (or 3.5% dissolved solids). Therefore, groundwater, even from a brine aquifer (see Chapter 6), will not pose any problems for bacterial growth.

In soil, water potential is used instead of water activity. It is defined as the difference in free energy between the system under study and a pool of pure water at the same temperature, and includes matrix and osmotic effects. The unit of measurement used is the *MPa*. As with water activity, this determines the amount of work that the cell must expend to obtain water. Generally, activity in soil is optimal at -0.01 MPa (or 30-90% of saturation) and decreases as the soil becomes either waterlogged near zero or desiccated at large, negative water potentials.[3] The in situ section of this chapter will show how this relates to percent moisture in soils.

When a microorganism grows in an environment with low water activity (high solute content or low moisture content), the cell must expend energy to extract water from solution. This usually results in a lower growth rate. This is most common with microorganisms growing in soil or exposed to the air. Below a certain water activity (0.60 to 0.70) microbial activity will cease, and unless the organism is resistant to desiccation it will die. Bacterial spores, and the sexual spores of fungi and algae are resistant to drying and can remain dormant for long periods of time. When water is reintroduced, microbial activity is reestablished.

pH

Microorganisms have ideal pH ranges that allow growth. Within these ranges, there is usually a defined pH optimum. Generally, the optimal pH for bacteria is between 6.5 and 7.5, which is close to the intracellular pH. A bacteria cell contains approximately 1000 enzymes and many are pH dependent.[3] Most natural environments have pH values between 5 and 9. Only a few species can grow at pH values of less than 2 or greater than 10.[8] In environments with pH values above or below the optimum value, bacteria are capable of maintaining an internal neutral pH by preventing H^+ ions from leaving the cell or by actively expelling H^+ as they enter. Once again, the most important factor with pH is to prevent major alterations in pH during the project.

Toxic Environments

Many factors can make an environment toxic to microorganisms. Physical agents such as high and low temperatures, sound, radiation, and hydrostatic pressure can influence microbial growth. Chemical agents such as heavy metals, halogens, and oxidants can inhibit microbial growth. This section will briefly discuss how these factors influence microbial growth.

As previously mentioned, high temperatures will inactivate enzymes, denature microbial proteins and nucleic acids. This will stop enzymatic reactions, weaken or rupture cell walls, and cause leakage of nucleic acids. Low temperatures will slow or stop cellular activity. Freezing prevents microbial growth but does not always kill the organism. When cells are subjected to freezing temperatures, their cytoplasm does not freeze as fast as the surrounding environment. The rate of temperature change will affect the formation of ice crystals. Ice crystals result in an increase in the concentration of solutes in the water left within the cell, effectively causing dehydration. Many organisms do not survive dehydration. Ice crystal formation can also damage intracellular components especially the plasma membrane, thus causing cellular death.[8]

Sound and radiation are not typically important factors with respect to environmental remediation. However, sound in the ultrasonic range can cause cavitation within cells disrupting cellular function.[12] Radiation in the ultraviolet and short x-rays will cause disruption of cellular activity. Ultraviolet light will transform cellular DNA, thus preventing successful replication. X-rays are absorbed and convert molecules and atoms into ions that can break molecular bonds.[12] Some of the new, enhanced vapor extraction systems use these types of radiation. The radiation effect on the natural bacteria will have to be evaluated when considering these new methods.

Hydrostatic pressure can affect microbial growth. Hydrostatic pressure affects the activity of most enzymes, protein synthesis, and membrane transport. Most bacteria isolated from shallow water or soil grow best at atmospheric pressure and are inhibited or killed at hydrostatic pressures of 200 to 600 atmospheres.[8]

Chemical agents such as heavy metals and halogens can disrupt cellular activities by interfering with protein function. Mercury ions combine with SH groups in proteins, silver ions will precipitate protein molecules, and iodine will iodinate proteins containing tyrosine residues preventing normal cellular function. The effects of various metals in soil has been described[13] and is affected by the concentration and pH of the soil. Oxidizing agents such as chlorine, ozone, and hydrogen peroxide oxidize cellular components thus destroying cellular integrity. Some in situ methods use hydrogen peroxide as an oxygen source. Oxidation and destruction of cells is possible with high concentrations.

Microbial Biodegradation of Xenobiotic Organic Compounds

The susceptibility of a xenobiotic compound to microbial degradation is determined by the ability of the microbial population to catalyze the reactions necessary to degrade the organics. Readily degradable compounds have existed on earth for millions of years, therefore, there are organisms that can mineralize these compounds. Industrial chemicals have been present on earth for a very short period on the evolutionary time scale. Many of these compounds are degradable, but many are persistent in the environment. Some xenobiotic compounds are very similar to natural compounds and bacteria will degrade them easily. Other xenobiotic compounds will require special biochemical pathways in order to undergo biochemical degradation.

A few definitions would be helpful here in order to understand different levels of biological reactions. Biodegradation means the biological transformation of an organic chemical to another form.[14] Biodegradation does not have to lead to complete mineralization. Mineralization is the complete oxidation of an organic compound to carbon dioxide. Recalcitrance is defined as the inherent resistance of a chemical to any degree of biodegradation, and persistence is defined to mean that a chemical fails to undergo biodegradation under a defined set of conditions.[15] This means that a chemical can be degradable but due to environmental conditions, the compound(s) may persist in the environment. With proper manipulation of the environmental conditions, biodegradation of these compounds can be demonstrated in laboratory treatability studies and the conditions transferred to field implementation.

As described above, microorganisms contain enzyme systems that are capable of cleaving C-C bonds of an organic compound. However, many xenobiotic compounds are not structurally capable of immediately entering microbial biochemical pathways and must be modified. Compounds such as alkanes, saturated ring structures, and unsubstituted benzene are biochemically inert and must be oxygenated before dehydrogenation reactions can occur.[14] Bacteria contain oxygenase enzymes that are capable of making molecular oxygen react with organic compounds and thus produce fragments that can then enter the normal metabolic pathways.

Gratuitous Biodegradation

Enzymes are typically described as proteins capable of catalyzing highly specific biochemical reactions. Enzymes are more specific to organic compound functional groups than to specific compounds. As Grady[14] described, an enzyme will not differentiate between a C-C bond in a benzene molecule and a C-C bond in a phenol molecule. The functional capability of enzymes depends on the specificity exhibited towards the organic compound. A major enzymatic mechanism used by bacteria to degrade xenobiotic compounds has been termed gratuitous biodegradation and includes existing enzymes capable of catalyzing a reaction towards a chemical substrate.

In order for gratuitous biodegradation to occur, the bacterial populations must be capable of inducing the requisite enzymes specific for the xenobiotic compound. Often times this occurs in response to similarities (structural or functional groups) with naturally occurring organic chemicals. For example, a bacteria is producing the enzymes for benzene degradation. Chlorobenzene is introduced and is not recognized by the bacteria (its presence will not induce an enzyme to be produced). However, the enzymes already produced for benzene will also catalyze the degradation of chlorobenzene.

The capability of bacterial populations to induce these enzymes depends on structural similarities and the extent of substitutions on the parent compound. Generally, as the number of substitutions increases, biodegradability decreases unless a natural inducer is present to permit synthesis of the required enzymes. To overcome potential enzymatic limitations, bacteria populations often produce a series of enzymes that work together to modify xenobiotic compounds. Each enzyme will modify the existing compound such that a different enzyme may be specific for the new compound and capable of degrading it further. Eventually, the original xenobiotic compound will not be present and the compound will resemble a natural organic compound and enter into normal metabolic pathways. This concept of functional pathways is more likely to be completed through the combined efforts of mixed communities of bacteria than by any single species.

Cometabolism

Cometabolism has recently been defined as "the transformation of a nongrowth substrate in the obligate presence of a growth substrate."[14] A nongrowth substrate cannot serve as the sole carbon source to provide energy to support metabolic processes. A second compound is required to support biological processes, allowing transformation of the nongrowth substrate. This requirement is added to make a distinction between cometabolism and gratuitous biodegradation.

Grady[14] presents an example to clarify the distinction between cometabolism and gratuitous biodegradation. Consider a situation in which a culture of cells includes enzymes that are capable of catalyzing degradation of a xenobiotic compound. Gratuitous biodegradation occurs if the xenobiotic compound added to the pregrown culture of cells transforms. (Remember, this is gratuitous biodegradation, and the culture could not grow and the transformation would eventually cease.) Cometabolism occurs when energy is required to complete transformation of the xenobiotic compounds. The xenobiotic compounds added to the pregrown culture, would not transform due to the inability of the culture to extract energy from the substrate. Only when an energy yielding substrate is added to the culture, would the transformation occur. Table 4-3 gives several examples of cometabolite compounds.

Microbial Communities

Complete mineralization of a xenobiotic compound may require more than one microorganism. No single bacteria within the mixed culture contains the complete genome (genetic makeup) of a mixed community. The microorganisms work together to complete the pathway from organic compound to carbon dioxide. These associations have been called consortia, syntrophic associations, and synergistic associations and communities.[14] We need to understand the importance of the community when we deal with actual remediations.

TABLE 4-3   Organic Chemicals Modified by Cometabolism

Acenaphthalene	Dodecane
Alkyl benzene sulfonate	Ethane
Anthracene	Ethene
Benzene	Ethylbenzene
bis(4-Chlorophenyl) acetic acid	Heptadecane
Butane	Hexadecane
1-Butene	4-Isopropyltoulene
cis-2-Butene	Limonene
trans-2-Butene	2-Methylanthracene
n-Butylbenzene	2-Methylnaphthalene
n-Butylcyclohexane	3-Methylphenanthrene
Carbon monoxide	Naphthalene
3-Chlorobenzoate	Octadecane
4-Chlorotoluene	Pentadecane
Cumene	Phenylcyclohexane
Cyclohexane	Propane
Cycloparaffins	Propene
p-Cymene	n-Probylbenzene
DDT	Retene
n-Decane	Tetradecane
1,2-Diethylbenzene	Thianaphthene
Diethyl ether	Toluene
9,10-Dimethylanthracene	2,4,5-Trichlorophenoxyacetate
1,3-Dimethylnaphthalene	Tridecane
2,3-Dimethylnaphthalene	1,2,4-Trimethylbenzene
1,6-Dimethylnaphthalene	Undecane
2,7-Dimethylnaphthalene	m-Xylene
	p-Xylene

Source: Dragun 1988.

Conversely, we need to understand the limitations of laboratory work with single organisms. This work does not represent the real world of degradation. Reviewing the strengths of the communities will also reveal the limitations of adding specialized bacteria that have been grown in the laboratory.

Community Interaction

Microbial communities are in a continuous state of flux and constantly adapting to their environment. Population dynamics, environmental conditions, and growth substrates continually change and impact complex interactions between microbial populations. Even though environmental disturbances can be modified by microorganisms, microbial ecosystems lack long-term stability and are continually adapting.[14] It is important to understand the complexities and interactions within an ecosystem to prevent failure when designing a biological system.

The existence of specific microbial interactions within communities is difficult to prove and this difficulty has prompted investigators to classify members of microbial communities on a functional basis. Organisms that degrade xenobiotic compounds have been divided into two groups: the primary utilizers and the secondary organisms.[14,15] The primary utilizers are those species capable of metabolizing the sole carbon and energy source provided to the system. The secondary organisms cannot use the major substrate and rely on the products generated by the primary utilizers.

Communities and Adaptation

Mixed communities have greater capacity to biodegrade xenobiotic compounds because of the greater genetic diversity of the population. Complete mineralization of xenobiotic compounds may rely on enzyme systems produced by multiple species. Community resistance to toxic stresses may also be greater because of the likelihood that an organism can detoxify the ecosystem.

Community adaptation is dependent upon the evolution of novel metabolic pathways. As described by Grady,[14] a bacterial cell considered in isolation has a relatively limited adaptive potential, and the adaption of a pure culture must come from mutations. Mutations are rare events. These mutations are generally responsible for enzymes that catalyze only slight modifications to the xenobiotic compound. An entire pathway can be formed through the cooperative effort of various populations. This is due to the greater probability that an enzyme system capable of gratuitous biodegradation exists within a larger gene pool. This genetic capability can then be transferred to organisms lacking the metabolic function which enhances the genetic diversity of the population. Through gene transfer, individual bacteria have access to a larger genetic pool allowing evolution of novel degradative pathways.

Genetic Transfer

Genes are transferred throughout bacterial communities by three mechanisms called conjugation, transformation, and transduction.[8,10,12,14,21] Conjugation appears to be the most important mechanism of gene transfer in the natural environment. Conjugation involves the transfer of DNA from one bacteria to another while the bacteria are temporarily joined. The DNA strands that are transferred are separate from the bacterial chromosomal DNA and are called plasmids.[8,10,12,21] Plasmids exist in cells as circular, double-stranded DNA, and are replicated during transfer from donor to recipient. Unlike chromosomal DNA which encodes for life sustaining processes, plasmid genes encode for processes that enhance growth or survival in a particular environment. Examples of functions that are encoded on plasmids include antibiotic resistance, heavy metal resistance, and certain xenobiotic degradation (i.e., toluene).[21]

Plasmids are not necessarily species specific, thus successful transfer of genetic material from many different species of the microbial community may occur.[21] Transposons, which are smaller DNA fragments, are also able to be transferred. These fragments incorporate into viral DNA, plasmids, and the chromosomal DNA of bacteria. Therefore, genes that have entered one cell type as transposons may eventually enter another cell type after being transported into it as a plasmid, minimizing transfer barriers among the community.[14] There are limitations to interspecies and intraspecies transfer of genetic material. These limitations are influenced by the environment, stability between the mating pair, and contact time between donor and recipient.[8,10,12,21]

Degradation Rate

Microbial degradation of organic material is generally described as the time necessary to transform the substrate from its original form to another form. The final form can be a structurally different compound or complete mineralization into carbon dioxide, water, oxygen, and other inorganic matter. Biodegradation rates can be measured by the loss of the original substrate, consumption of oxygen, or the evolution of carbon dioxide or methane.

Biodegradation rates can be described by two reaction rates.[3,13,16] These rates are called zero-order and first-order.

Reactions that transform the substrate and are unaffected by changes in the substrate concentration are called zero-order kinetics.[3,13,16] In these cases, the reaction rate is determined by some factor other than the substrate concentration. For example, if the cell density is so great that the quantity of substrate is insufficient to support a significant increase in cells, the kinetics of the disappearance of organic compounds present is zero-order (linear with time).

First-order reactions occur when the biodegradation rate of a substrate is proportional to the concentration of the substrate. This reaction has been demonstrated in the literature for biodegradation of many organic compounds.[13] First-order kinetics described for a single bacterial species includes two patterns. In the first pattern there is no appreciable increase in cell numbers. The bacteria have reached a threshold or the initial cell number is too large, relative to the quantity of organic compound, to permit an appreciable increase in bacteria. At constant biomass or severely limiting substrate levels, the degradation rate is proportional to the concentration of residual substrate, which falls off continually. In the second pattern, few active cells are initially present, and the chemical concentration is above any threshold concentration that may limit the degradation rate. Under these conditions the bacteria will grow, but at a rate that falls constantly with the diminishing and always limiting substrate concentration.

In the real world, the reaction rates can change during the project. For these situations, hyperbolic rate law or the Monod equation describes the growth of the microorganism population as a function of the substrate level over a range of concentrations. This function is dependent on substrate concentration and the growth rate of the microorganisms, and it is particularly useful when the initial concentration is in the mixed-order region. Figure 4-5 summarizes the various growth rate models as disappearance of organic chemicals.

One final subject needs to be discussed under degradation rate, low concentrations. We normally think of treatment process efficiency in terms of percent removal. In Chapter 3, we discussed air strippers and related their performance to the percent removal of a specific organic. A packed tower air stripper required a certain height of packing in order to remove 90% of TCE. While the original tower design was based upon the effluent requirements, the final system removed 90% no matter what the influent concentration was. The design was, therefore, based on the maximum influent concentration.

Biological systems cannot always be evaluated based upon percent removal. At low concentrations, diffusion of the compounds to the cell surfaces may not be sufficient for growth or maintenance of the microbial populations. This concentration which occurs when biological activity is reduced, is called the threshold and is controlled by

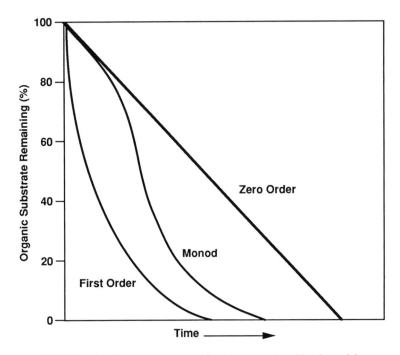

FIGURE 4-5. Disappearance curves based upon various kinetic models.

the substrate concentration and diffusivity. No matter what the influent concentration, the biological reaction may not continue past this threshold value. Therefore, percent removal is not always a good measurement to describe the performance of a biological system. This is especially true with the low concentrations we normally find with groundwater contaminations.

This natural limitation in the bacterial reaction rates forces us to develop additional methods to measure the performance of biological reactors. Two other methods for evaluation could be influent and effluent concentrations. Biological systems require a minimum amount of food in order to replace the bacteria that are lost to decay and washout. A biological reactor efficiency can be measured by the influent concentration required in order to maintain a viable population.

Effluent concentration is also an important measurement. Will the reactor design achieve an effluent concentration equal to the natural threshold level of the organic compound? Will the reactor be able to achieve concentrations below the threshold level? We will use all of these methods in the Biological Reactor section. Finally, concentration thresholds may also help to explain the persistence of low concentrations of biodegradable compounds in the environment. Once a remediation reaches the low ppb level in the aquifer, even degradable compounds may be very slow to reach final concentrations. Bacteria may have limited affect on these compounds at these concentrations.

Degradative Mechanisms

We have discussed how the degradation of different organic compounds require different enzymes, and that different microorganisms are required for specific degradation. However, all degradation is related. One degradation pathway is central to all microorganisms. The pathway where organic compounds are oxidized to release energy or used as a substrate to form cellular components is called the Tricarboxylic Acid Cycle (TCA). The key compound in the TCA cycle is acetyl-CoA (acetyl radical coupled to coenzyme A) derived from pyruvate. Acetyl-CoA binds to oxalacetate to form citric acid that continues through the cycle producing carbon dioxide, NADH (ATP formation), and two intermediates (α-ketoglutarate and oxalacetate) used in amino acid synthesis Figure 4-6.

Bacteria degrade compounds because they recognize them as food, not because they are doing us a favor. The degradation of xenobiotic

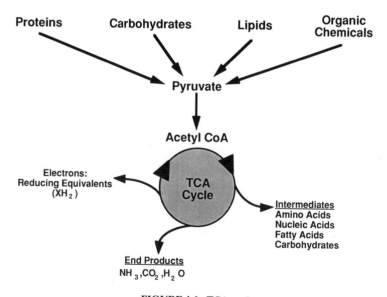

FIGURE 4-6. TCA cycle.

compounds is dependent upon the capability of deriving energy or biosynthetic fragments from the compound. The objective of the enzymes that we are trying to induce is to get the xenobiotic compound into the TCA cycle. If the enzymes are induced and the reactions proceed, the products are then readily metabolized by the organism's biochemical pathways. All microorganisms use the TCA cycle. Only the pathways to the TCA cycle differ.

Many different pathways are available, depending on the compound that is being degraded. The following sections give some examples of the degradation mechanisms of several types of organic compounds.

Aliphatic Hydrocarbons

There are several microorganisms that are capable of degrading aliphatic hydrocarbons.[8,12,13,17] These reactions are strictly aerobic. These compounds represent a large percentage of the compounds found in petroleum hydrocarbons.

Figure 4-7 is a summary of the degradation of n-octane. The initial step involves the reaction of molecular oxygen with one of the carbon molecules of the hydrocarbon. Monoxygenases are most generally involved and bind oxygen to the terminal methyl group or the second carbon. Subsequent reactions form fatty acids that can be incorporated in the cell or can be further oxidized by β-oxidation (fatty acid oxidation).[8]

Cyclic Hydrocarbons

Degradation of cyclic hydrocarbons is similar to the degradation mechanism of aliphatic hydrocarbons. The products of degradation are further degraded by β-oxidation.[13] Degradation of cyclic hydrocarbons with functional groups becomes more complicated because more than one reaction is available. Figure 4-8 shows the pathway for degradation of cyclohexane to an aliphatic, and Figure 4-9 shows the degradation of cyclohexane to carboxylic acid.

Aromatic Hydrocarbons

Aromatic hydrocarbon degradation involves dioxygenases. The product formed is catechol, a dihydroxybenzene which is broken down, leading to either acetyl-CoA or TCA intermediates. Several aromatic compounds possessing one or more six-carbon rings such as benzoic acid, ethyl benzene, phthalic acid, phenanthrene, naphthalene, anthracene, toluene, phenol, and naphthol follow similar degradative

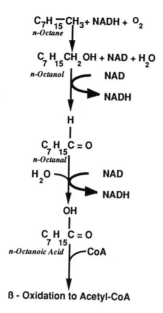

C_7H_{15}—CH_3 + NADH + O_2
n-Octane

$C_7H_{15}CH_2OH$ + NAD + H_2O
n-Octanol

NAD

NADH

H
|
$C_7H_{15}C = O$
n-Octanal

H_2O

NAD

NADH

OH
|
$C_7H_{15}C = O$
n-Octanoic Acid ⌐CoA

ß - Oxidation to Acetyl-CoA

$$CH_3{-}(CH_2)_4{-}CH_2{-}CH_2{-}\overset{\displaystyle O}{\overset{\|}{C}}{-}CoA$$

$$CH_3{-}\overset{\displaystyle O}{\overset{\|}{C}}{-}CoA \longrightarrow \text{TCA Cycle}$$
Acetyl CoA

CoA

$$CH_3{-}(CH_2)_2{-}CH_2{-}CH_2{-}\overset{\displaystyle O}{\overset{\|}{C}}{-}CoA$$

$$CH_3{-}\overset{\displaystyle O}{\overset{\|}{C}}{-}CoA \longrightarrow \text{TCA Cycle}$$
Acetyl CoA

CoA

$$CH_3{-}CH_2{-}CH_2{-}\overset{\displaystyle O}{\overset{\|}{C}}{-}CoA$$

$$CH_3{-}\overset{\displaystyle O}{\overset{\|}{C}}{-}CoA \longrightarrow \text{TCA Cycle}$$
Acetyl CoA

$$CH_3{-}CH_2{-}CoA \longrightarrow \text{TCA Cycle}$$

FIGURE 4-7. Degradation of N-octane.

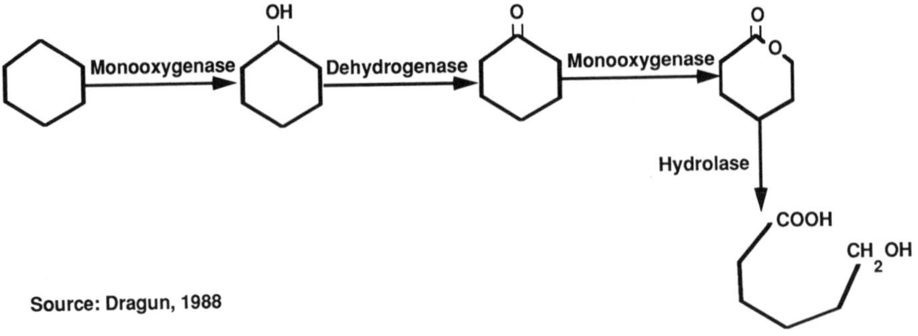

Source: Dragun, 1988

FIGURE 4-8. Degradation of cyclohexane to an aliphatic hydrocarbon.

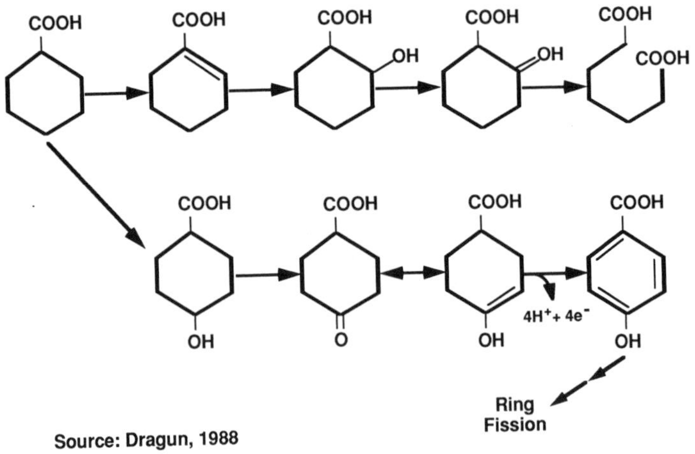

Source: Dragun, 1988

FIGURE 4-9. Degradation of cyclohexane to caboxylic acid.

pathways.[18] Figure 4-10 shows the degradation pathway for an aromatic hydrocarbon.

Halogenated Hydrocarbons

There are many pathways available for the degradation of halogenated hydrocarbons. Some of the compounds degrade under anaerobic conditions. Some require a co-metabolite for degradation. Many investigators have reported the mechanism of halogenated hydrocarbon transformation. Figure 4-11 is a compilation of several investigators research into degradation of common halogenated hydrocarbon pollutants.[19,22,23,24,25,26,27,28]

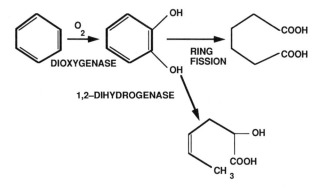

FIGURE 4-10. Degradation of an aromatic hydrocarbon.

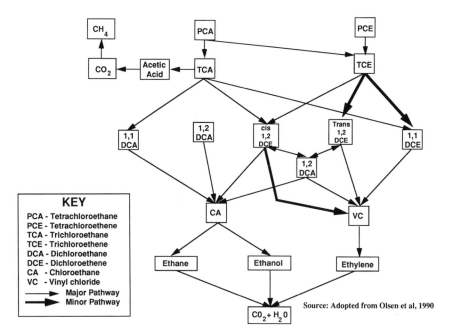

FIGURE 4-11. Transformations of chlorinated aliphatic hydrocarbons.

BIOLOGICAL REACTORS FOR CONTAMINATED WATER

Now that we have a better understanding of how biochemical reactions take place, and how to degrade xenobiotic compounds, we need to explain how to use these mechanisms in the real world. We will first discuss this for above ground reactors (water and soil) and then discuss in place or in situ biological treatment. The first point to make about biological reactors is to establish the difference between a biological reaction and a biological reactor. The biological reaction is bacteria producing requisite enzymes to use a specific contaminant as a food and energy source. In other words, is the compound degradable? The biological reactor is designed to maximize the biological reaction in an economic manner. The reactor design solves specific problems that are encountered in the contaminated streams that are being considered.

We must remember that the biological designs that are available today were originally designed to solve problems associated with wastewater not groundwater. Therefore, we can not simply say that "activated sludge" will not work on our groundwater problem and think that it represents all biological reactors. If the reaction can occur, then we have to use a biological reactor design that addresses the problems associated with groundwater when we analyze the applicability of bacteria for our groundwater problem.

There are two basic problems to be solved in the design of a biological reactor. The first is the contact between the bacteria and the organic contaminants. The second is the oxygen transfer to the bacteria. We can compare the various biological reactor designs that are presently available by considering how they solve these two problems. Other criteria will also be used to establish the advantages and disadvantages of specific reactor designs for groundwater, but bacterial contact and oxygen transfer are the two functions that are common to all reactor designs.

Bacterial contact is more than simply mixing the bacteria with the organic contaminants. The goal of the biological reaction is to destroy a maximum amount of the organics, and leave a minimum concentration of the contaminants remaining in the water. To achieve these goals, the bacteria must be in contact with the organics and have extended periods of time to complete the biochemical reactions. In other words, the bacteria must have a long residence time in the reactor. Figure 4-12 shows the relationship between the effluent organic concentration and the residence time of the bacteria.

Oxygen transfer does not affect the performance of the reactor design as long as a minimum oxygen concentration is maintained. Oxygen transfer is mainly related to the cost of biological treatment. As shown in Chapter 2, energy for oxygen transfer is the main operat-

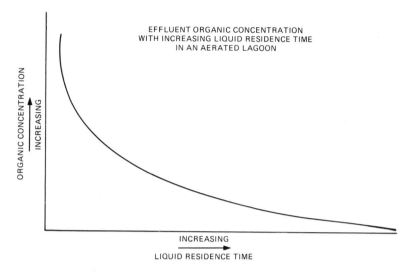

EFFLUENT ORGANIC CONCENTRATION
WITH INCREASING LIQUID RESIDENCE TIME
IN AN AERATED LAGOON

FIGURE 4-12. Effluent organic concentrations with increasing bacterial residence time.

ing cost of a biological reactor other than manpower. For large wastewater treatment systems, oxygen transfer is a major consideration in the economics of the design. Also as shown in Chapter 2, this function will have a less dramatic affect in groundwater designs. However, the original reason certain reactors were designed in a particular manner was to solve an oxygen transfer problem. In order to understand the differences between the reactors, we must understand the original design criteria.

Biological reactors can be separated into three main groups: suspended growth reactors, fixed film reactors, and miscellaneous designs. With suspended growth reactors, the bacteria are grown in the water and are intimately mixed with the organics in the water. Fixed film systems grow bacteria on an inert support media. The water containing the organics passes over the film of bacteria. A miscellaneous category is needed because there have been many special designs developed during the last several years. All of the designs have advantages and disadvantages.

Suspended Growth Reactors

The easiest unit operation to implement for a groundwater treatment system is an aerated lagoon or basin. An existing pond or tank can be used for the reactor. In some cases portable swimming pools have been used for the aeration tank. Figure 4-13 shows the configuration for an aerated lagoon. The contaminated groundwater enters the aerated vessel. Bacteria in the reactor degrade the organics and create new bacteria. The liquid residence time in the reactor, which is equal to the bacterial resident time, must be sufficient for the bacteria to reproduce before they exit with the water.

Oxygen is supplied to the tank with a surface aerator or air diffusers. Sufficient power must be supplied to provide an adequate oxygen concentration, 2 mg/l, and/or to keep the tank completely mixed. With low residence time reactors, oxygen supply is usually the controlling factor. Mixing is usually the controlling factor for long residence time reactors. Two days is the lowest residence time that should be used to maintain low effluent organic concentrations. This is about the time that it takes for the bacteria to reproduce and replace the cell washout with the effluent. At lower residence times, the bacteria are washed out of the reactor and insufficient bacteria remain for efficient removal of organics. Longer residence times will only be limited by total flow and the cost of power to mix the tank. The main operating cost (other than personnel) for any aerobic, biological treatment system is the power requirement for oxygen supply or mixing. This will limit the size of the reactor.

Two problems arise with the use of an aerated lagoon design. First, the degree of treatment is limited by the limited bacteria residence time. Lagoons can only be expected to remove 50 to 70% of the biodegradable organics. Second, the bacteria that are created in the reaction will be in the water when it leaves the reactor. A clarifier can be added to the system to remove the solids, but bacteria grown in an

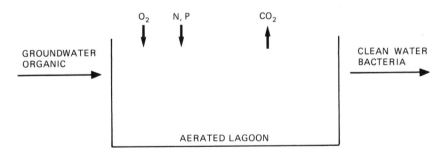

FIGURE 4-13. Aerated lagoon.

aerated lagoon do not settle readily. These limitations make the lagoon design suitable only for use in some in situ treatment situations. Even as part of in situ treatment design, a clarifier is added to limit the amount of solids that are returned to the soil. Direct discharge, and even discharge to another treatment system, require more thorough treatment.

These problems can be solved by separating the liquid residence time from the bacterial residence time. Figure 4-14 shows that by adding a clarifier to remove the bacteria solids from the water stream and returning it to the aerated reactor, the bacterial residence time is now separate. From Figure 4-14, the liquid residence time (R_L) is:

$$R_L = V/Q \tag{4-1}$$

The bacterial residence time (R_B) is:

$$R_B = \frac{X \times V}{Q_W \times X_R + (Q - Q_W)X_E} \tag{4-2}$$

The bacterial residence time is controlled by the loss of bacteria due to wasting the settled bacteria from the clarifier, and by the uncontrolled loss of bacteria over the top of the clarifier. When bacteria are

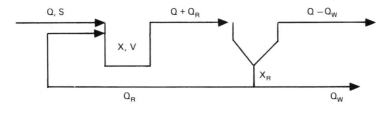

Q = Flow
Q_R = Recycle Flow
Q_W = Sludge Wastage Flow
X = Mixed Liquor Suspended Solids (MLSS)
X_R = Clarifier Underflow Solids Concentration
X_E = Effluent Solids Concentration
V = Volume of Aeration Basin
S = Organic Concentration

FIGURE 4-14. Activated sludge.

returned to the aeration tank, the process is called, activated sludge, Figure 4-14.

Activated sludge is the most widely used method of biological treatment in the wastewater treatment field. The basic advantages are: the process produces low effluent concentrations, the system can treat many organics at the same time, the same equipment can be used for a variety of influent conditions. The main disadvantages are: the cost of manpower to keep the system adjusted to the influent conditions, the relative cost of oxygen transfer compared to fixed film systems, and the critical need to keep the bacteria at a growth stage in which their settling characteristics are at a maximum.

Chapter 2 presented a design example of an activated sludge system over the life-cycle of a groundwater cleanup. The activated sludge process can remove 85 to 95% of the biodegradable organics from the influent, and 99% + of specific organic compounds (depending on influent concentrations). Effluent concentrations of 10 to 30 mg/l of biochemical oxygen demand (BOD), a general measurement of biodegradable organics in water, can be expected with a well-run system. Effluent concentrations of specific compounds will depend on that particular compound. For example, effluent phenol concentrations from an activated sludge system can be as low as 0.01 mg/l.

Certain compounds, i.e., sugars and alcohols, will degrade very quickly in a biological system. Other compounds require longer contact times with the bacteria in order to degrade. The easier a compound can be assimilated by the bacteria, the faster, and more efficiently the bacteria can turn that compound into new bacteria. When compounds are hard to degrade, they are persistent, and the bacteria need more residence time in order to replace their lost numbers. Another way to look at this is that the bacteria must first remove the easily degradable organics before they are willing to produce the enzymes necessary to degrade the refractory compounds. They must be a little hungry before they go after these organics. In the design of the treatment plant, this can be represented by the food to microorganism ratio, F/M. Figure 4-14, the formula would be:

$$F/M = Q \times S/V \times X \qquad (4\text{-}3)$$

In activated sludge, all of the compounds are being degraded at the same time in the completely mixed tank. The reader can use either of these models, bacteria residence time (sludge age), or food to micro-organism ratio, to help understand how to accomplish this concurrent degradation. One of the main advantages of the activated sludge process is that the bacterial residence time and the F/M can be controlled to accommodate the degradation of a variety of compounds even if they have different degradation rates.

The life-cycle design example in Chapter 2 showed that this versatility can be maintained over a variety of influent conditions. With innovative engineering, the process can be made to work over a large percentage of the life-cycle of the project.

The main problem with the activated sludge design is the critical need to keep the bacteria in a form in which they settle readily. If the bacteria do not settle properly, the clarifier will not be able to remove them from the water stream. If the bacteria are not separated from the water stream and returned to the aeration basin, the whole process fails. This does not mean that the bacteria have lost their ability to degrade organics. The bacteria will still be able to degrade the compounds. But, they will no longer be able to separate the bacterial residence time from the liquid residence time. The end result is just an aerated lagoon and the corresponding removal rates.

In order to maintain the settling properties, two things are necessary: the environment in which the bacteria grow must not have any major changes, and the bacteria must be grown at the proper sludge age that promotes flocculation. In groundwater treatment, the influent has very little variation, on a day to day basis. There is normally no need for equalization as is true in wastewater treatment. The main problems when using activated sludge with groundwater are the life-cycle concentration and the growing of the bacteria in their flocculant stage during the entire project. The example in Chapter 2 showed one way to overcome these problems. It is important for the design engineer to design the activated sludge treatment system to be operable over as much of the life-cycle as possible. Even with a good design, the activated sludge system will still require a relatively high level of operator attention to ensure that the system is operating in the correct manner.

There are two other types of equipment design that use suspended growth bacteria: extended aeration and contact stabilization. Neither of these designs have a particular benefit for groundwater treatment, but both have been widely used in wastewater treatment. Activated sludge, extended aeration, and contact stabilization are shown together in Figure 4-15. As can be seen in that figure, extended aeration is very similar to activated sludge. The only difference between the two designs is that the reaction chamber (aeration basin) is larger in the extended aeration. This increases (extends) the aeration of the bacteria. Extended aeration is more stable (the larger tanks serve as internal equalization), and it produces less waste sludge.

Most "packaged plants" that can be purchased from vendors are extended aeration designs. Therefore, the reader has a good chance of encountering the proposed application of this design on groundwater. The extended aeration design will have the same limitations on groundwater as did the activated sludge design.

Contact stabilization is widely used for high concentrations of easily degradable organics. As can be seen in Figure 4-15, the waste comes into contact with the bacteria in a small aeration tank. The

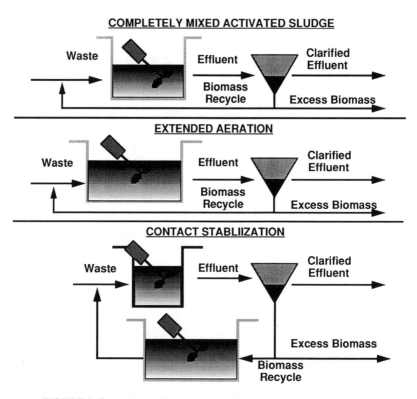

FIGURE 4-15. Activated sludge, extended aeration, contact stabilization.

bacteria quickly assimilate the organics without digesting them. The bacteria, with the organics, are removed in a clarifier and sent to a large aeration tank. The bacteria digest (stabilize) the organics in this tank. After the bacteria have digested the organics, they return to the contact tank to pick up more organics, and the cycle starts again. The main purpose of this design is to save space and subsequent operating costs. It is unlikely that the reader will encounter this design in a groundwater treatment system. However, it may be mentioned in feasibility studies and so the reader should be aware of its basic design. The advantages and disadvantages of suspended growth designs are summarized in Table 4-4.

Another way to use bacteria in an above ground treatment system is to set up a fixed film biological unit. In fixed film systems, large surface area, inert support media is placed in the reactor. Bacteria naturally attach and grow on any surface provided to them, Figure 4-16. The contaminated water enters the tank and forms a thin film over the attached bacteria. The contaminants transfer into the attached bacteria. The bacteria degrades the organics and the waste byproducts (CO_2, H_2O) transfer back to the water film. Oxygen transfers from the atmosphere, through the water film and to the bacterial fixed film. There are two important advantages of the fixed film systems: the bacteria are maintained in a high concentration without the need of a clarifier, and oxygen can be supplied at lower costs.

A third advantage is the general ease of operation. A fixed film system does not require the operator attention that an activated sludge system does. Bacteria will grow attached to the medium and remove organics from the water over a wide range of operating conditions. When there are too many bacteria in the system, the

TABLE 4-4   Suspended Growth Systems

Advantages	Disadvantages
Intimate contact between biomass and waste	Completely dependent on clarifier performance
Several methods available for adjusting performance	High operation attention required
Very low concentration of specific organics in effluent	
Large scale system relatively inexpensive	

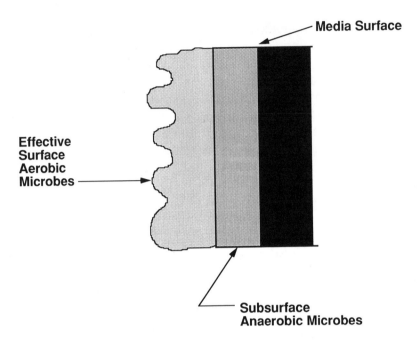

FIGURE 4-16. Fixed film bacterial growth.

bacteria will sluff off the medium and leave the reactor with the water (a clarifier can be used remove the solids before final discharge).

There are two main types of fixed film reactors. Trickling filters and rotating biological contactors (RBC). Figure 4-17 shows a trickling filter design. The contaminated water is pumped to the top of the reactor, and distributed over the medium. The water is broken up into thin films and trickles down through the medium. The contaminants transfer into the bacterial film and degrade. Oxygen transfers through the thin film of water and to the bacteria. Waste by-products (i.e., carbon dioxide) transfer out through the thin film of water into the atmosphere.

Several types of inert support media can be used in a trickling filter. Originally, small (3 to 5 inches) rocks were used to support the bacterial population. However, rocks were limited to a small bacterial mass in the system because of low surface area per unit volume, and low oxygen transfer capacity because of low void space. Plastic media have replaced rocks in recent years. The two main categories of plastic media are dumped packing and stacked packing. Dumped packing is the same type of plastic medium used in packed tower air

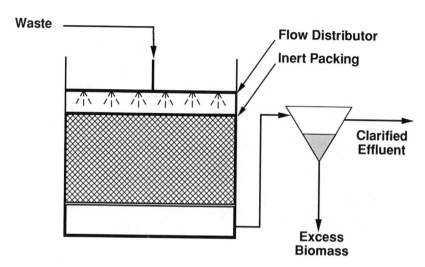

FIGURE 4-17. Trickling filter.

strippers. Stacked medium comes in large bricks and is shown in Figure 4-18. Dumped packing is usually used as a replacement for rocks in existing trickling filters and small new systems. Stacked packing is usually applied to large systems.

Figure 4-19 shows an RBC design. In this system, the water enters one end of the tank. The medium first rotates down into the water. The contaminants, once again, transfer to the bacteria. The medium then rotates up into the atmosphere and a thin film of water forms on the medium, and the oxygen transfers through the film of water and to the bacteria. RBCs are probably the most energy efficient oxygen transfer method for biological system.

There are several technical disadvantages with the fixed film reactors. Fixed film reactors are plug flow reactors. The water comes in at one end, passes by the bacterial film, and exits at the other end of the reactor. The influent end of the reactor sees the high influent concentration of the contaminant. In completely mixed reactors, the influent is immediately mixed with the contents of the tank. The influent concentration may be toxic, or pockets of high concentration material may be found as the groundwater is recovered from the ground. The bacteria in the fixed film reactor will be subjected to the full concentration. Recycling of the effluent water can be used to minimize this effect, but it also adds to the cost of operation.

Another problem with fixed film reactors is that they will not

FIGURE 4-18. Stacked packing. (Courtesy of Munters Corp.)

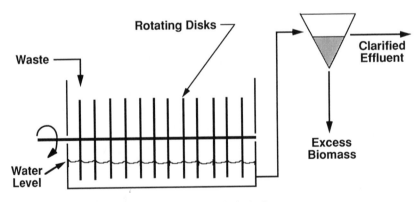

FIGURE 4-19. Rotating biological contactor.

remove as high a percentage of the influent contaminant as will an activated sludge system. Specific chemical removal is very important in groundwater treatment. General removal of organics will be important, depending upon the final disposal of the water. The design engineer can expect 75 to 90% BOD removal and 85 to 95% removal of a specific organic. As discussed before, the lower the influent con-

TABLE 4-5   Fixed Film Reactors

Advantages	Disadvantages
Low operator attention	Plug flow
Retention of slow growing bacterial populations	Limited operation at high influent concentrations
Low cost oxygen transfer	Hard to adjust operation
Resistent to shock loads	

centration, the less percentage of removal can be expected. Table 4-5 summarizes the advantages and disadvantages of the fixed film systems.

Submerged Fixed Film Reactors

A relatively new biological design is a combination of suspended growth and fixed film reactors designs. These can be referred to as submerged fixed film reactors. Figure 4-20 shows a submerged fixed film unit.

In these units, the medium is placed in the reactor tank and the water level is maintained above the height of the plastic medium. The bacteria grow on the plastic medium as in a fixed film system, however the water is in intimate contact with the film as opposed to passing through in thin films.

There are two main ways in which the submerged fixed film design can be utilized. The first way has been used for many years in

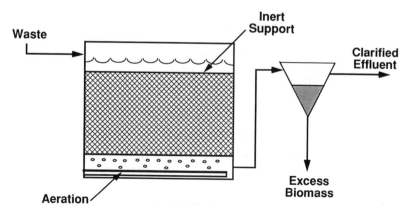

FIGURE 4-20. Submerged fixed film.

wastewater treatment.[29] An example of a commercial unit is shown in Figure 4-21. The reactor is designed for completely mixed operation and for high concentrations of organic influent. Air is released below the medium, and travels up through it. The air pushes water in front of it as it rises, creating an air lift pumping action. With sufficient air, the biological reactor tank is completely mixed. This design can handle from 50 to 5,000 mg/l influent organic content.

The main advantages of the submerged fixed film unit in this design are ease of operation and high quality performance. The submerged fixed film can perform as well as an activated sludge unit, however, it is not dependant upon a clarifier for maintaining the bacteria in the reaction tank. This allows for a large variety of operating conditions and for low operator attention. This reactor design combines the advantages of the suspended growth systems and the advantages of the fixed film systems without the disadvantages of either. The main disadvantages of the submerged fixed film design are high cost of oxygen transfer and lack of scale ability. Due to the nature of the design, there is a natural height limitation to the tank and therefore oxygen cannot be released at an optimum depth. In addition, diffused aeration is one of the more costly oxygen transfer methods. The second problem is the scaling of the unit. Suspended growth and fixed film units cost relatively less as the systems get larger. Because the tank and the medium both get larger in direct relationship to the size of the system, the submerged fixed film does not have large cost advantages for larger systems.

Neither of these disadvantages has a large affect on applications on groundwater. First, oxygen transfer is a small part of the total cost of a biological treatment system in groundwater systems. Second, most groundwater treatment systems are relatively small and the cost advantage of large-scale systems do not apply.

A second manner in which the submerged fixed film units can be applied to groundwater is with a low concentration design. Submerged fixed film can be designed to treat influent concentrations as low as 1 to 20 mg/l. This is very important for groundwater applications because high concentrations (greater than 50 mg/l) are rarely found in groundwater, and when they are, life-cycle design considerations bring the concentration below 50 mg/l in a short period of time.

Figure 4-22 shows the low concentration design of a submerged fixed film unit. The basic design is the same as the original submerged

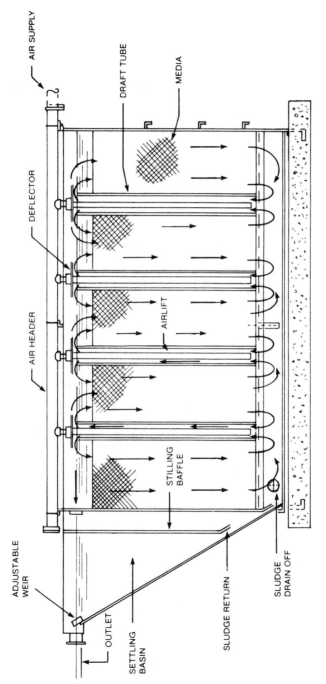

FIGURE 4-21. Fixed activated sludge treatment. (Courtesy of Smith & Loveless, Inc.)

Labels within figure: AIR SUPPLY, DRAFT TUBE, MEDIA, DEFLECTOR, AIR HEADER, AIRLIFT, STILLING BAFFLE, ADJUSTABLE WEIR, OUTLET, SETTLING BASIN, SLUDGE RETURN, SLUDGE DRAIN OFF

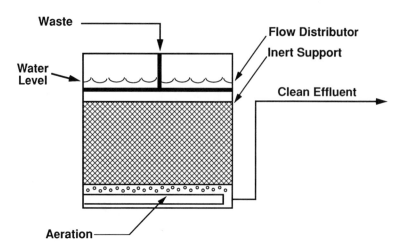

FIGURE 4-22. Low concentration submerged fixed film.

fixed film design. A plastic medium is submerged below the water level in the reactor tank. The low concentration reactor uses small amounts of air and a plug-flow pattern. The water enters the top of the tank and is distributed across the medium. The water flows down through the medium and exits the bottom through a collection system. The air is released below the medium. Very small amounts of air are used because of the low requirement for oxygen demand in a low concentration reactor and the need to maintain a nonmixed state in the reactor.

Even under these conditions, the low concentration of influent organics is not sufficient to maintain an active biological population in the reactor. Normally, influent concentrations of less than 50 mg/l will result in bacteria decay faster than new bacteria can grow to replace the old bacteria. Therefore, the low concentration reactor must operate in a decay mode, not in the normal growth mode of biological treatment systems. In this decay mode, bacteria are grown on the fixed film from a synthetic feed source. Once the bacteria have established a full population, the synthetic feed is removed and the low concentration influent is conveyed through the system. Under these conditions, the bacteria slowly decay. With the proper design and operation, the decay period can last between six months to one year before regrowth is required.

The low concentration reactors are no longer in the laboratory or pilot-plant stage. The author has personally designed over 20 units that are currently operating at full-scale in the field. Compounds that have been treated in this reactor design have ranged from acetone and MEK to benzene and chlorobenzene. Such reactors are also currently working on tetrahydrofuran and t-butanol.

Miscellaneous Treatment Systems

There are several other types of designs that have been used on organic contaminants. Several of the most popular designs in recent years have been the powder activated carbon treatment (PACT)™ units, units which sequence batch reactors, and fluidized bed reactors. All of these systems have advantages and disadvantages, and the reader will have to review their strengths and weaknesses for their particular requirements.

The PACT™ system is shown in Figure 4-23. The system is basically an activated sludge treatment system with powder activated carbon maintained in the reactor. The combination of powder activated carbon and active bacteria strengthens the removal capabilities of the treatment system. The powder activated carbon can remove slow degrading or nondegrading organic material from the water while the bacteria can attach to the powdered activated carbon and consume the organics that have absorbed to the carbon. The system has been widely used on hazardous organic waste. While it has mainly been used in the wastewater field, it is now being applied to the groundwater field. The main advantage of the system is that it can treat a large variety of organic compounds. The main disadvantage of the system

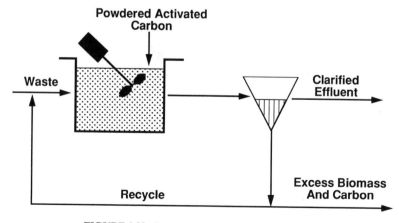

FIGURE 4-23. Powdered activated carbon treatment.

is that it is basically an activated sludge design, and will suffer from the same limitations that such systems usually encounter.

The sequencing batch reactor has also been widely applied to industrial waste during the last several years. Figure 4-24 shows the operational sequence of a sequencing batch reactor. The main steps to the operation are:

1. Fill the reaction tank with the contaminated water while maintaining full aeration,
2. Once the tank is full, complete digestion of the organics by the bacteria,
3. Stop aeration and subsequent mixing and allow the bacteria to settle,
4. Decant the clean water and discharge.
5. Start the cycle again.

The sequencing batch reactor design basically uses two tanks in parallel. While one reactor is accepting water, the other reactor is going through the subsequent steps of digestion, settling, and decanting. The reactors switch back and forth to maintain a constant influent flow. The advantages of the sequencing batch reactor are simplicity of operation and the variety of influent conditions that can be accommodated. The main disadvantage for groundwater would be operation under low concentrations of influent organics.

A fluidized bed reactor is basically a submerged fixed film type of

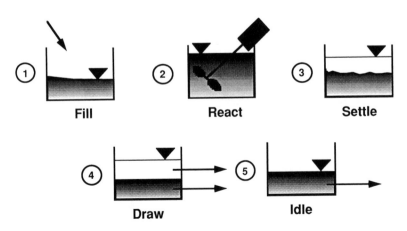

FIGURE 4-24. Sequencing batch reactor treatment stages.

design. In this design the support medium is very small. Water and air flow in an up flow pattern through the medium, fluidizing the bed. Typical media are sand, activated carbon, glass beads, and other small particles. Small packing allows for very high concentrations of bacteria that can be maintained within the reactor. This design has advantages for high concentrations of organics but has not widely been applied to full-scale installations.

Anaerobic reactors are another possible design that can be applied to groundwater situations. All of the laboratory work that is presently occurring in this area shows that bacteria can be used to degrade chlorinated hydrocarbons. Once these reactions are well understood, and the required environment defined, then full-scale reactors will be possible. Currently, there are several anaerobic reactor designs available. Their main purpose is for the treatment of high concentration wastewater. The food and beverage industries are the main areas of application. This chapter has limited its discussion on anaerobic biochemical reactions. This is because of the limited full scale data available from groundwater applications. As the laboratory work reaches the field and the knowledge becomes more practical, the readers should include anaerobic reactors as part of their treatment alternatives.

Several other types of biological reactors are available from vendors. New types of designs are constantly coming on the market. With the exception of the low concentration submerged fixed film design, there are no other current units that are specifically designed for groundwater situations. Once again, it is very important to understand that just because activated sludge is not a viable design for the groundwater, this does not mean that biological treatment is not possible. Any new design for groundwater treatment will have to address the specific problems of the groundwater design. The main problem areas are life-cycle design, low concentration of organics, portability, and low operator attention. The designer will have to understand the advantages and disadvantages of each biological unit, and use those designs to their best advantage during the life-cycle of the project.

BIOLOGICAL REACTORS FOR CONTAMINATED SOILS

While the main subject of this book is the treatment of contaminated groundwater, several closely related topics are also covered. These topics are concerned with the remediation of the contaminated soils and/or sediments, which are frequently present at a site where the groundwater quality has deteriorated. Biological treatment of organic contaminants adsorbed to soil particles will be covered in this section. Bioremediation of soils may be performed either in situ or ex situ. In situ bioremediation, the biological treatment of contaminated soils and ground water without excavation, is discussed later in this chapter. Some sites are not suited to in situ treatment because of the hydro-geological conditions at the site or to the characteristics of the wastes present. At such sites, soils may be remediated using ex situ treatment. Ex situ bioremediation is the term for the biological treatment of excavated soils. This section is concerned with the various types of bioreactors used for ex situ treatment.

Ex situ biotreatment reactors for soil remediation fall into two main categories: solid-phase treatment and slurry-phase treatment. Solid-phase biotreatment relies on principles applied in agriculture and in the biocycling of natural compounds. Solid-phase process options include engineered-landfarm treatment, soil-pile treatment, and composting. In slurry-phase treatment, contaminated soils or sludges are maintained as an aqueous slurry.

There are advantages and disadvantages to each of these designs. However, all of the reactors follow the basic concepts of any biological reactor. Their main purposes are to maintain an environment in which the organics and bacteria can react under optimum conditions. The bacterial requirements of environment (pH, temperature, etc.) and nutrient addition (O_2, NH_3, PO_4, etc.) all must be met by the reactor. The solid phase reactors have one more requirement. They must also maintain the proper moisture content.

Engineered-Landfarm Treatment

Landfarming is the practice of spreading wastes over the ground surface to enhance natural microbial degradation of contaminants. Landfarming was the first method used for bioremediation of soils and sludges and has been successfully applied by the petroleum industry for the managed disposal of petroleum-refinery wastes for decades.[30] Engineered landfarming refers to a landfarm treatment system that has been designed for containment, and possibly treatment, of soil leachate and volatile organic emissions.

A typical engineered-landfarm layout is depicted in Figure 4-25. A clean sand drainage layer is placed upon an impermeable liner (double liner or concrete bottom, as required) and equipped with perforated drainage piping to create a leachate collection system. Contaminated soils are placed on top of the sand. The landfarm includes a spray irrigation system for application of nutrients and inoculum, if desired, and for controlling the soil moisture content. These items can also be added by the same soil mixing equipment that is used for oxygen transfer. A greenhouse top and air management system may also be included if containment of volatile emissions is required. The contaminated soil is tilled regularly (e.g., daily, every other day, or weekly depending upon oxygen requirements) to promote homogenization of the soil and increase the oxygen available to the indigenous microorganisms.

As shown in Figure 4-25, the collected leachate drains to a lined sump. The leachate may be pumped to an onsite liquid-phase bioreactor for treatment, with treated leachate used as a source of microbial

CONTAINED ABOVE GROUND SOILS TREATMENT

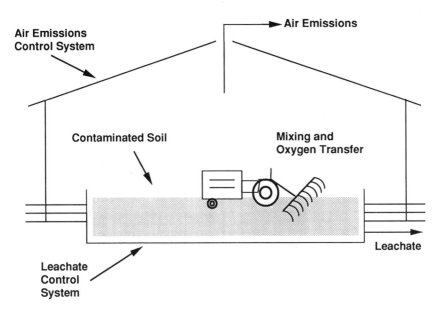

FIGURE 4-25. Engineered landfarm.

inocula. After nutrients have been added, the treated leachate may be reapplied to the landfarm.

Engineered landfarming is commonly used today for biotreatment of soil contaminated with petroleum and wood-preserving wastes. At a pilot scale landfarm constructed at an oil-gasification site, successful bioremediation of approximately 6,000 cubic yards of coal-tar contaminated soil was achieved. The contaminated soil was placed in a bed to a 2 ft depth and regularly tilled and irrigated. Results achieved in a four-month treatment period included a 73% reduction in BTX concentrations, a 36% reduction in the concentration of oil and grease, and an 86% reduction in the concentration of total PAHs. Two-ring and three-ring PAH concentrations decreased by 92%; four-ring PAH concentrations decreased by 80%; and five-ring PAH concentrations decreased by 65%.[31]

Bioremediation of soil contaminated with wood-preserving wastes was successfully demonstrated in another pilot study. The contaminants monitored were pentachlorophenol (PCP) and creosote, which is a complex mixture containing a number of polynuclear aromatic hydrocarbons (PAHs). During a five-month period, the concentration of PCP was reduced by 95%, while reductions in PAH concentrations ranging from 50 to 75% were achieved.[32]

Soil-Pile Treatment

There are two types of soil pile reactors. One delivers oxygen and nutrients by water movement through the soil. The other mixes the nutrients with the soil when the pile is created, and delivers oxygen by air movement through the soil.

The first soil-pile treatment is identical to engineered-landfarm treatment, except that no tilling of the soil is involved. In soil-pile treatment, contaminated soil is spread on a lined treatment bed equipped with a drainage collection system. An irrigation system allows for the constant flow of a solution containing nutrients and inoculum. The collected irrigation stream drains to a sump, from which it may be conveyed to a liquid-phase bioreactor. The bioreactor effluent would be channeled back through the irrigation system. The soil-pile system may be totally enclosed if volatile-emissions control is necessary.

Like engineered-landfarm treatment, the first soil-pile treatment has been successfully applied to soil contaminated with petroleum and wood-preserving wastes. At one wood-preserving site, a soil-pile reactor was constructed within an existing RCRA impoundment area. Approximately 1200 cubic yards of sludge and contaminated soils were mixed with an equal volume of native soils and spread in a six-inch layer in the reactor. The soil pile was irrigated daily to maintain the desired moisture content within the bed.

The contaminants monitored were benzene-extractable (BE) hydrocarbons and 16 PAHs. BE hydrocarbon concentrations were reduced by 69% over the first year of operation, with most of the reduction occurring during the first four months (i.e., May through September). Average removal rates of 95% were achieved in the first year in the cases of both two-ring and three-ring PAHs. The average total PAH reduction rate was 90%, while the rate for four- and five-ring PAHs was 72%. As was true of the BE hydrocarbons, the greatest PAH reductions were achieved in the warm weather months. These results were expected since the rate of biological degradation would be greatest during warmer weather. Further contaminant reductions achieved over the winter months were slight.[33,34,35]

The second soil pile design is more versatile. The water based design is limited in size and oxygen transfer because of its reliance on water movement. The air reactor can be larger and handle higher concentrations than can the water soil pile.

The air soil pile is constructed in a similar fashion to the water soil pile. An impermeable layer is placed first on the ground. Then the soil (mixed with nutrients and required inoculum) is placed directly on the liner. There is no need for a leachate collection system since no water is involved. Air pipes are placed in the soil as the pile is created. The air pipes are used to deliver and/or collect air. The spacing of the pipes is dependent upon the permeability of the soil. The pipes are connected to the vacuum side of the blower. The exhaust air can be treated if required. The pile can be any size, but is usually limited to a maximum 12-20 foot height and width. A plastic liner is also placed on top of the pile to keep rainwater out and to help with the proper air movement patterns through the pile. Figure 4-26 shows an air soil pile reactor.

Composting

Composting is another modification of solid-phase bioremediation. Composting is an established technology for the management of

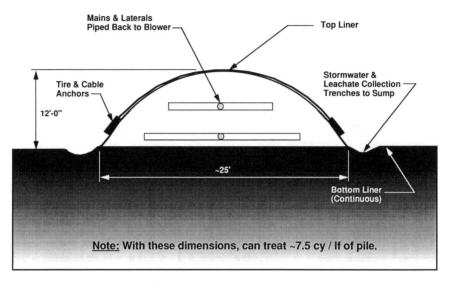

FIGURE 4-26. Soil pile reactor.

municipal wastewater treatment sludges and is an emerging technology for the treatment of hazardous wastes. Biodegradation of organic compounds occurs within a compost matrix consisting of contaminated soils and/or sludges mixed with organic carbon sources and bulking agents such as straw, bark, or wood chips. The compost piles, which are three to six feet in height, are placed on lined treatment beds. The microbial degradation of contaminants is enhanced by aeration and maintenance of optimum temperature, moisture content, and nutrient levels within the piles.

There are three types of compost-treatment systems: open-windrow systems, static-windrow systems, and in-vessel systems.[36] The compost is stacked in elongated piles (i.e., windrows) in open- and static-windrow systems. In-vessel composting involves placing the compost inside an enclosed reactor. Aeration of open-windrow systems is accomplished by mechanically turning the windrows. Compost piles are built on top of a grid of perforated pipes in static-windrow and in-vessel applications and they are aerated using forced-air systems. All three designs may include systems for leachate collection and containment of volatile emissions.

Decomposition releases energy in the form of heat that remains trapped within the compost matrix, resulting in the self-heating characteristic of composting.[38] Composting of hazardous waste may not cause the same increase of temperature. The addition of bulking agents enhances the porosity of the compost matrix, facilitating mixing and oxygen transfer. The bulking agents are generally screened out after the completion of composting and may then be added to a subsequent batch of contaminated soil to be composted. Once again, for hazardous waste, the process may have to be modified.

The process is suitable for highly contaminated soils or sludges, poorly textured wastes, and in applications where temperature is critical to the sustained treatment process.[38] Composting has been successfully used for treatment of petroleum sludges and crop residues contaminated with pesticides or other toxic organics.[39] Two field-scale studies were performed to investigate the feasibility of using composting to remediate explosives and propellant contaminated sediments at two Army ammunition plants. Contaminants of concern at one site included 2,4,6-trinitrotoluene (TNT), hexahydro-1,3,5-trinitro-1,3,5-triazine (RDX), and octahydro-1,3,5,7-tetranitro-1,3,5,7-tetraazocine (HMX). The contaminant of concern at the other site was nitrocellulose (NC).

Four compost piles of approximately 12 cubic yards each were constructed at each site. The piles were placed on 8-inch thick concrete pads, which drained to a sump. The compost mixture consisted of cow or horse manure, straw, alfalfa, horse feed, and contaminated sediment. The compost piles were moistened with water at the time they were placed on the concrete pads. Each compost pile contained a network of perforated pipes through which air was drawn to aerate the piles.

The compost period lasted 22 weeks at the first site. During that time the following results were achieved: reductions in TNT concentrations ranging from 99.6 to 99.9%; reductions in RDX concentrations ranging from 94.8 to 99.1%; and reductions in HMX concentrations ranging from 86.9 to 96.5%. At the second site, reductions in NC concentrations ranging from 91 to 98% were achieved during a 14-week composting period.[37]

Slurry-Phase Biotreatment

Slurry-phase biotreatment may be performed either in bioreactor vessels or in lined lagoons, but the basic features include aeration equipment, mechanical mixing, and sometimes, an emissions control syste .[40] Therefore, depending on the setting, slurry-phase bioremediation may be compared to the activated sludge process or aerated-lagoon treatment. In either application, soil or sludge and nutrient-adjusted water are combined to form an aqueous slurry which is biodegraded. Mixing must be sufficient to keep the solids in suspension, and oxygen must be supplied throughout the slurry matrix to promote aerobic microbial activity. Bioslurry reactors are operated to maximize mass transfer rates and contact time between the contaminants and microorganisms.[41]

Pretreatment for removal of oversize material is required to achieve effective treatment of contaminated soil in a bioslurry reactor. The first step in the treatment process is to slurry the soil or sludge to be treated. Afterward, it is passed through a trammel screen to remove gravel and debris larger than 0.25 inch in diameter. More water may then be added to obtain the desired slurry density prior to bioslurry treatment. Maximum treatment efficiencies are generally obtained with soil slurries containing 30 to 50% dry solids by weight, although the difficulty of maintaining the solids in suspension limits the acceptable slurry solids-content range to 20 to 30% for some bioslurry reactors.[38,42]

The EIMCO Biolift™ reactor shown in Figure 4-27 is a bioslurry reactor designed to minimize energy requirements by means of a dual-drive system. The dual drive permits the mixing impeller and rake arms to be operated at two different speeds. Fine-bubble diffusers are mounted on the rake mechanism, which turns slowly (e.g., 2 to 4 RPM), generating a rotating curtain of fine bubbles. The diffused air bubbles keep most of the fine particles in suspension and create the necessary turbulence to enhance the mass transfer of oxygen and nutrients to the microorganisms. Mixing is controlled by the impeller drive, which turns at higher speeds (e.g., 20 to 30 RPM), producing a downward slurry flow in the center of the vessel and an upward flow along the tank walls. The rotational speed of the impeller is insufficient to keep coarser particles in suspension. Coarser particles that have settled to the bottom of the vessel are raked to a central airlift and pumped to the top. The use of an airlift to maintain a completely-mixed flow regime is the energy-saving feature of this design. Other units, which utilize high-speed mixers to maintain coarse materials in suspension, are reportedly more energy-intensive to operate.[43]

Three pilot-scale demonstrations of slurry-phase treatment, each lasting 60 days, were performed at a site contaminated with oil-

EIMCO BIOLIFT™ REACTOR

SCHEMATIC

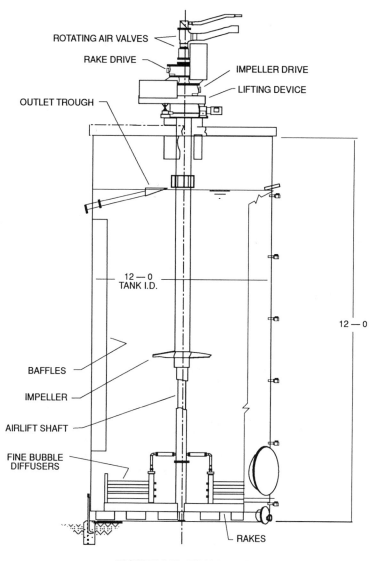

ROTATING AIR VALVES

RAKE DRIVE

IMPELLER DRIVE

LIFTING DEVICE

OUTLET TROUGH

12 — 0
TANK I.D.

12 — 0

BAFFLES

IMPELLER

AIRLIFT SHAFT

FINE BUBBLE
DIFFUSERS

RAKES

FIGURE 4-27. EIMCO biolift.

181

refining wastes. The contaminants of concern were PAHs, oil, and grease. The treatment vessels ranged from a 17,000-gallon reactor to a 750,000-gallon aerated lagoon. The reactor solids loadings ranged from 5 to 30% (i.e., the slurries consisted of 5 to 30% dry solids by weight). Data pertaining to the oil and grease treatment results were not reported. However, reductions in total PAH concentrations ranging from 76 to 92% and in carcinogenic PAH (i.e., five-ring and six-ring PAHs) concentrations ranging from 25 to 89% were achieved. Although the experimenters expected to achieve the highest removal efficiencies in the reactors with the lowest solids loadings, this hypothesis was not supported. In fact, the highest total and carcinogenic PAH reductions were obtained in the lagoon with 30% solids loading.[40]

Slurry systems have an economic and technical advantage when the contaminated solids already contain a high moisture content. Lagoon bottoms are a prime example. In addition to the physical advantages of not having to add water, the reactors actually perform two functions. The first function is the normal degradation of the contaminants. The second function is a reduction of volume. This is accomplished through direct degradation, and by the breaking of the emulsion with the subsequent release of soils and water. Even when the soils require further treatment, this second function can still make the slurry treatment economically worthwhile.

Comparison of the Costs of
Ex Situ Bioremediation Technologies

Per-ton costs for the four ex situ bioremediation technologies are presented in Table 4-6. All unit costs tabulated assume that the waste treated is contaminated soil having a unit weight of 110 pounds per cubic foot. As shown in Table 4-6, engineered-landfarm treatment and composting are attractive because of their low costs, which are in the ranges of $35 to $80 per ton and $40 to $100 per ton, respectively.[38,44,45] Soil-pile treatment costs are in the $90 to $100 per-ton range, while slurry-phase treatment costs range from $80 to $150 per ton.[42,48,49,50] All of the ex situ bioremediation technology costs compared favorably to the costs shown for off-site landfill disposal (i.e., $200 to $300 per ton) and off-site incineration (i.e., $300 to $2,000 per ton). The landfill-disposal and incineration costs shown do not include transportation costs, which are substantial. For example, transportation

TABLE 4-6   Treatment Reactors Costs for Solid Phase Biological

Process Options	Cost
Off-site disposal in permitted hazardous waste landfill	$200-$300/ton plus transportation costs
Off-site incineration in permitted facility	$300-$1200/ton plus transportation costs
Engineered land-farm treatment	$35-$100/ton
Soil pile treatment	$50-$100/ton
Composting	$50-$70/ton
Bioslurry reactor treatment	$80-$150/ton

costs for a 20,000 pound truckload of contaminated soil hauled by a qualified hazardous-materials transporter range from $2.50 to $3.50 per loaded mile.

APPLYING BIOLOGICAL REACTORS

Biological treatment has become an important technology for groundwater remediation. Biological treatment is an inexpensive method when designed correctly. It is a destruction method. The organics do not exist after being removed by the system. Many organics can be treated at the same time. The bacteria that are left behind will continue to clean up the soil and the aquifer after the treatment system is shut down. It is the natural system. Nature uses bacteria to break down complicated organic compounds (cellulose from trees and plants is a persistent compound) in order to recycle the components. This is part of the carbon cycle. However, there are limitations to the process, and the design engineer should make sure that biological treatment is applicable to the particular situation. Biological reactors must run 24 hours per day, 7 days a week. You cannot turn them on and off as with physical/chemical systems. Biological treatment is not applicable to all organic compounds. It should not be applied when the treated water is to be used for final consumption by humans or animals unless the water is carefully treated afterward to remove all of the bacteria. Most biological systems require two to six weeks to start up. Bacteria are difficult to apply to short term projects other than in situ treatment.

There are several steps that the design engineer should take before deciding on biological treatment for the cleanup of a particular organic compound:

1. Literature search for data on the degradability of the compound,
2. Analyze media for general organic parameters—BOD, COD, TOC,
3. Conduct treatability studies,
4. Select the biological process to be applied, and
5. Construct and operate an on-site biological treatment pilot plant.

The design engineer must first decide if bacteria can degrade the compounds that have been found in the groundwater and soil. This can be accomplished by any of the first three steps listed above. Probably, all three steps should be used, except in cases involving easily degradable compounds like sugars and alcohols.

There has been a considerable amount of work performed on the degradability of different organic compounds. The problem is that, for the persistent compounds, the literature does not always agree. Table 4-7 lists a compilation of several literature sources on the degradability of 30 specific organic compounds. As can be seen from Table 4-7, the more substitutions (halogen or other inorganic groups replacing the hydrogen molecule attached to the carbon) on an organic, the more difficult the organic is to degrade. As long as one literature source finds the compound degradable, the design engineer should proceed with the evaluation. Even if no previous successes can be found, the engineer may apply certain degradation enhancement techniques as discussed in the first section of this chapter during the treatability studies to promote degradation of the compound.

Before the treatability study, general organic parameter analyses should be performed. Biochemical oxygen demand (BOD), chemical oxygen demand (COD), and total organic carbon (TOC) tests should be completed. These tests will tell the engineer the total amount of organic material in the groundwater, not just the specific compound. Total organic content in soils is more difficult to determine. TOC or total petroleum hydrocarbon (TPH) tests have been used to estimate the total amount of contaminants in soils. The design of the treatment system is based on the total amount of degradable organic material. The ratio of the results from these tests will give the design engineer some idea about the degradability of the organic compound in the

TABLE 4-7   Disappearance or Biodegradation Potential for Specific Organic Compounds

Compound	Biodegradability	Disappearance Rate	Environmental Condition	Reference
Acetone	D			
Benzene	D	43% in 7d	si, nmf	1
		110d half life	scf, sdw	2
		68d half life	sgw, fo	3
		48d half life	gwi nm	3
		20–90% in 80d	si, nmf	4
		100% in 434d	sgw, fo	4
		>99% in 120w	si, nmf	5
		66–100%	as	19
		>60%	ai	19
		84–96%	al	19
		30 mg.d^{-1}	gw	4
Bromodichloromethane	P,D	0%	al	20
		35% in 7d	scf, sdw	1
Bromoform	P,D	8% in 7d	scf, sdw	1
		>99% in 2d	cfc, bm	6
Carbon Tetrachloride	P,D	84% in 7d	scf, sdw	1
		>99% in 2d	cfc, bm	6
		>98%	as	20
Chlorobenzene	D	60% in 7d	scf, sdw	1
		37d half life	sgw, fo	2
		<3.8% in 1w	si, nmf	9
		0.2–1.9% in 1w	si, nmf	10
		37d half life	aq	2

(continued)

185

TABLE 4-7   Disappearance or Biodegradation Potential for Specific Organic Compounds—*Continued*

Compound	Biodegradability	Disappearance Rate	Environmental Condition	Reference
Chloroform	P,D	5-100%	as	20
		36-86%	al	20
		48% in 7d	scf, sdw	1
		68% in 27d	swi, nmf	7
		96% in 2d	cfc, bm	6
		3% in 5d	si, nmf	8
2-Chlorophenol	D	85% in 7d	scf, sdw	1
Dichlorobenzene	P,D	46% in 7d	scf, sdw	1
		110d half life	sgw, fo	2
1,1-Dichloroethane	P,D	40% in 7d	scf, sdw	1
		<1.2-<2.6%/w	si, nmf	10
1,2-Dichloroethane	P,D	23% in 7d	scf, sdw	1
		>99% in 2d	cfc, bm	6
		>60d half life	U	24
1,1-Dichloroethylene	P,D	62% in 7d	scf, sdw	1
		68% in 4d	swi, nmm	14
		92% in 40w	si, nmf	5
		110d half life	swi, nmf	18
cis-1,2-Dichloroethylene	P,D	49% in 7d	scf, sdw	1
		100% in 50h	swi, nmm	14
		100% in 16w	si, nmf	5
		140d half life	swi, nmf	18
trans-1,2-Dichloroethylene	P,D	54% in 7d	scf, sdw	1
		100% in 50h	swi, nmm	14
		92% in 40w	si, nmf	5
		139d half life	swi, nmf	18

Compound				
Ethylbenzene	D	85% in 7d	scf, sdw	1
		100% in 12d	bgw, nmf	11
		37d half life	sgw, fo	2
		>99% in 120w	si, nmf	5
		100% in 192h	sp, nmf	12
		37d half life	aq	2
Hexachlorobenzene	P,R	39% in 7d	scf, sdw	1
Methyl ethyl ketone	D			
Methylene chloride	D	100% in 7d	scf, sdw	1
		34–72%	as	20
		65%	al	20
Naphthalene	D	100% in 7d	scf, sdw	1
		100% in 9d	bgw, nmf	11
		110d half life	sgw, fo	2
		100% in 1w	si, naf	17
		100% in 192h	sp, nmf	12
		1–14d half life	sm	28
		62%	as	19
		>62%	ai	19
		>29%	al	19
Pentachlorophenol	P,D	21–1087d half life	sm, nmf	28
		82–96% in 64d	sm, ph	29
		98% in 2 mo	sm, ph	30
		88–91% in 6.5w	fs, ph	31
		>95% in 28–35d	An, SL	32
Phenol	D	97% in 7d	scf, sdw	1
		98.5 mg COD/g/h	bss, as	16

(continued)

187

TABLE 4-7 Disappearance or Biodegradation Potential for Specific Organic Compounds—*Continued*

Compound	Biodegradability	Disappearance Rate	Environmental Condition	Reference
Tetrachloroethylene	P,D	38% in 7d	scf, sdw	1
		0% in 190h	swi, nmm	14
		300d half life	sgw, fo	13
		78–99.98% in 2–4d	cfc, nmm	15
		86% in 2d	cfc, bm	6
		68% in 21d	swi, nmf	7
		0.9–1.8%/w	si, nmf	10
Toluene	D	100% in 7d	scf, sdw	1
		100% in 10d	bgw, nmf	11
		37d half life	sgw, fo	2
		39d half life	sgw, fo	3
		37d half life	gwi, nmf	3
		100% in 30–80d	si, nmf	4
		100% in 80d	sgw, fo	4
		>99% in 120w	si, nmf	5
		>93%/w	si, nmf	9
		0.9–3.2%/w	si, nmf	10
		100% in 192h	sp, nmf	12
1,1,1-Trichloroethane	P,D	17yr half life	U	25
		230d half life	U	13
		26% in 7d	scf, sdw	1
		300d half life	sgw, fo	13
		98% in 2d	cfc, bm	6
		<1.1–<3.2%/w	si, nmf	10
1,1,2-Trichloroethane	P,D	3% in 7d	scf, sdw	1
		24d half life	U	24

Trichloroethylene	P,D	230d half life	U	13
		33d half life	U	18
		43d half life	U	24
		51% in 7d	scf, sdw	1
		69% in 4d	swi, nmm	14
		300d half life	sgw, fo	13
		<3.5%/w	si, nmf	9
		89% in 40w	si, nmf	5
		<1.2-<2.3%/w	si, nmf	10
Vinyl chloride	P,D	100% in 23d	swi, nmm	14
o-Xylene	D	100% in 12d	bgw, nmf	11
		11d half life	sgw, fo	2
		32d half life	sgw, fo	3
		31d half life	gwi, nmf	3
		100% in 25-60d	si, nmf	4
		100% in <300d	sgw, fo	4
		>99% in 120w	si, nmf	5
		100% in 192h	sp, nmf	12

Many of these citations were compiled from Dragun (1988), and Pitter and Chudoba (1990).

Notes:

A = Aerobic
An = Anaerobic
D = Degradable
De = Denitrification (anoxic conditions)
LS = Lab study
M = Mixed culture
P = Persistent
PC = Pure culture used as an inoculum
R = Recalcitrant
S = Soil used as an inoculum
SA = Soil or aquifer material used as inoculum
SL = Anaerobic sewage sludge added as inoculum

(continued)

U = Unknown conditions
ai = Activated sludge (industrial)
al = Aerated lagoon
aq = Aquatic
as = Activated sludge (sewage)
bgw = Batch test using groundwater
bm = Bacterial inoculum produced in a methanogenic environment
bss = Batch test using distilled water, dissolved salts, and the organic chemical as the sole carbon source
cfc = Continuous-flow, fixed film laboratory study using glass bead columns
d = day(s)
fo = Estimation based on field observation
fs = Field study
gw = Ground water aquifer system
m = Methanogenic conditions
mb = Mycobacterium sp. added as inoculum
naf = Natural acclimated microbial flora
nmf = Natural microbial flora used as inoculum
nmm = Natural microbial flora under methanogenic conditions
p = Pseudomonas sp. used as an inoculum
ph = Inoculated with phanerochaete sp.
scf = Static-culture flask biodegradation test, original culture
sdw = Settled domestic wastewater utilized as microbial inoculum
sgw = Naturally-occurring soil groundwater system
si = Soil incubation study
sm = Soil microcosom study
sp = Soil percolation study
swi = Soil/water or soil/sediment incubation study
w = Week(s)
x = Xanthobacter
y = Year(s)

References for Table 4-7
1. Tabak, H. H., Quave, S. A., Mashni, C. I., and Barth, E. F. Biodegradability studies with organic priority pollutant compounds. *Journal Water Pollution Control Federation* 53:1503-1518 (1981).
2. Zoeteman, B. C. J., De Greef, E., and Brinkmann, F. J.J. Persistency of organic contaminants in groundwater, lessons from soil pollution incidents in the Netherlands. *The Science of the Total Environment* 21:187-202 (1981).

3. Barker, J. F. and Patrick, G. C. Natural attenuation of aromatic hydrocarbons in a shallow sand aquifer. In *Proceedings of the NWWA/API Conference on Petroleum Hydrocarbons and Organic Chemicals in Groundwater— Prevention, Detection, and Restoration.* November 13-15, 1985, Houston, TX. Dublin, OH: National Water Well Association (1985).

4. Barker, J. F., Patrick, G. C., and Major, D. Natural attenuation of aromatic hydrocarbons in a shallow sand aquifer. *Ground Water Monitoring Review* 7:64-71 (1987).

5. Wilson, B. H., Smith, G. B., and Rees, J. F. Biotransformations of selected alkylbenzenes and halogenated aliphatic hydrocarbons in methanogenic aquifer material: A microcosom study. *Environ. Sci. Tech.* 20:997-1002 (1986).

6. Bouwer, E. J. and McCarty, P. L. Transformations of 1-and 2-carbon halogenated aliphatic organic compounds under methanogenic conditions. *Applied and Environmental Microbiology* 45:1286-1294 (1983).

7. Parsons, F., Wood, P. R., and DeMarco, J. Transformations of tetrachloroethene and trichloroethene in microcosoms and groundwater *Journal AWWA* 76:56-59 (1984).

8. Strand, S. E. and Shippert, L. Oxidation of chloroform in an aerobic soil exposed to natural gas. *Applied and Environmental Microbiology* 52:203-205 (1986).

9. Wilson, J. T., McNabb, J. F., Balkwill, D. L., and Ghiorse, W. C. Enumeration and characterization of bacteria indigenous to a shallow water-table aquifer. *Ground Water* 21:134-142 (1983).

10. Wilson, J. T., McNabb, J. F., Wilson, B. H., and Noonan, M. J. Biotransformation of selected organic pollutants in groundwater. *Developments in Industrial Microbiology* 24:225-233 (1982).

11. Kappeler, Th. and Wuhrmann, L. Microbial degradation of the water soluble fraction of gas-oil—II. Bioassays with pure strains. *Water Research* 12:335-342 (1978).

12. Kappeler, Th. and Wuhrmann, K. Microbial degradation of the water-soluble fraction of gas oil—I. *Water Research* 12:327-333 (1978).

13. Roberts, P. V., Schreiner, J. E., and Hopkins, G. D. Field study of organic water quality changes during groundwater recharge in the Palo Alto baylands. *Water Research* 16:1025-1035 (1982).

14. Fogel, M. M., Taddeo, A. R., and Fogel, S. Biodegradation of chlorinated ethenes by a methane-utilizing mixed culture. *Applied and Environmental Microbiology* 51:720-724 (1986).

15. Vogel, T. M. and McCarty, P. L. Biotransformation of tetrachloroethylene to trichloroethylene, dichloroethylene, vinyl chloride, and carbon dioxide under methanogenic conditions. *Applied and Environmental Microbiology* 49:1080-1083 (1985).

16. Pitter, P. Determination of biological degradability of organic substances. *Water Research* 10:231-235 (1976).

17. Wilson, J. T., McNabb, J. F., Cochran, J. W., Wang, T. H., Tomson, M. B., and Bedient, P. B. Influence of microbial adaptation on the fate of organic pollutants in groundwater. *Environmental Toxicology and Chemistry* 4:721-726 (1985).

18. Barrio-Lage, G., Parsons, F. Z., Nassar, R. J., and Lorenzo, P. A. Sequential dehalogenation of chlorinated ethenes. *Environ. Sci. Technol.* 20:96-99 (1986).

19. Patterson, J. W. and Kodukala, P. S. Biodegradation of hazardous organic pollutants. *Chem. Eng. Prog.* 77:48 (1981).

20. Richards, D. J. and Shieh, W. K. Biological fate of organic priority pollutants in the aquatic environment. *Water Research* 20:1077 (1986).

21. Vogel, T. M. and McCarty, P. L. Transformations of halogenated aliphatic compounds. *Env. Sci. Technol.* 21:722-736 (1987).

22. Reinkek, W. and Knackmuss, H. J. *Nature* 277:385-386 (1979).

23. Bouwer, E. J. and McCarty, P. L. Modeling of trace organics biotransformation in the subsurface. *Ground Water* 22:433 (1984).

24. Wood, P. R. and Lang, R. E., Payan, in *Groundwater Quality* C. H. Ward, W. Giger, P. L. McCarty, Eds., Wiley: New York, (1985).

(continued)

25. Vogel, T. M. and McCarty, P. L. *Environ. Sci. Technol.* 12:1208-1213 (1987).
26. Pitter, P. Biodegradability of some aromatic cycloaliphatic and aliphatic compounds by activated sludge. *Vodni Hospod. B,* 25:321 (1975) (in Czech).
27. Pitter, P. Relations between the structure of organic compounds and their biological degradability. *Chem. Prum.* 26:21 (1976) (in Czech).
28. Characterization and Laboratory Soil Treatability Studies for Creosote and Pentachlorophenol Sludges and Contaminated Soil. EPA: Washington, D.C., 1988, EPA/600/2-88/055.
29. Lamar, R. T., Larsen. M. J., and Kirk, T. K. Sensitivity to and degradation of pentachlorophenol by phanerochaete spp.. *Applied and Environmental Microbiology* 56:3519-3526 (1990).
30. Lamar, R. T., Glaser, J. A., and Kirk, T. K. Fate of pentachlorophenol (PCP) in sterile soils inoculated with the white-rot basidiomycete phanerochaete chrysosporium: Mineralization, volatilization and depletion of PCP. *Soil Biol. Biochem.* 4:433-440 (1990).
31. Lamar, R. T. and Dietrich, D. M. In situ depletion of pentachlorophenol from contaminated soil by Phanerochaete spp. *Applied and Environmental Microbiology* 56:3093-3100 (1990).
32. Mikesell, M. D. and Boyd, S. A. Enhancement of pentachlorophenol degradation in soil through induced anaerobiosis and bioaugmentation with anaerobic sewage sludge. *Env. Sci. Technol.* 22:1411-1414 (1988).
33. Baben, L. and Vaishnav, D. D., Prediction of biodegradability for selected organic chemicals. *J. Ind. Microbiol.* 2:107 (1987).
34. Vogel, T. M. and McCarty, P. L. Submitted for publication in *Environ. Sci. Technol.* (1987).
35. Hallen, R. T., Pyne, J. W., Jr., and Molton, P. M. Extended abstract. *Annual Meeting of the American Chemical Society* (1986).
36. Parsons, F., Wood, P. R., and DeMarco, J. *J. Amer. Water Works Assoc.* 76(5):56-59 (1984).
37. Reinkek, W. and Knackmuss, H. J. *Nature* 277:385-386 (1979).
38. Kobayashi, H. J. and Rittmann, B. E. Microbial removal of hazardous organic compounds. *Env. Sci. Technol.* 16:170A-183A (1982).
39. McCarty, P. L., Rittmann, B. E., and Bouwer, E. J. Microbial processes affecting chemical transformations. In G. Bitton and C. P. Gerba (Eds.). *Groundwater Pollution Microbiology* 89-115. J. Wiley & Sons, New York, 1984.
40. Schwarzenbach, R. P., Giger, W., Hoehn, E., and Schneider, J. K. Behavior of organic compounds during infiltration of river water to groundwater-field studies. *Environ. Sci. Technol.* 17:472-479 (1983).

groundwater. The ratio between the test results should be about: 1/2-3/1, for BOD/COD/TOC. Low values for BOD may indicate nondegradable organic material present. Low values for the TOC may indicate that inorganic oxygen demand is present.

Treatability studies should almost always be completed before setting up a full-scale treatment system. These studies will not be directly scalable to the full-scale system, but will give an accurate picture of the ability of the bacteria to degrade the various organics in the groundwater and soil. The studies will also show if there is something unexpected in the water or soil that will be toxic to the bacteria.

Treatability studies can also be used to try different sources of bacteria on the groundwater and soil. Bacteria can be obtained from various wastewater treatment plants in the geographic area of the clean up. A wastewater treatment plant that is already degrading the organic contaminants found in the groundwater and soil is the preferred source of bacteria. A plant treating structurally similar compounds would also be a good seed source; i.e., the groundwater has phenol in it and a local biological treatment plant treats benzene. A soil sample from the surface where the original spill occurred is also a good source of bacteria. It has been the experience of the author that most soils contain adequate bacterial populations for degrading contaminants. Therefore, soil reactors usually do not need seed sources of bacteria.

When an unusual compound must be degraded, samples of bacteria can be shipped in limited quantities from treatment plants and other spill cleanups across the country that have treated similar compounds. There are also commercial bacterial products that can be purchased from various companies. These products are available off-the-shelf. While there is a variety of products available that can degrade a wide range of organic compounds, the products do not degrade any compounds that natural bacteria cannot degrade. The main advantage of the commercial bacteria is convenience. A high concentration of stable bacteria can be purchased and shipped to the project site where the bacteria can be used when needed. When comparing the cost of bacteria to the total cost of the project, the cost of commercial bacteria is about 5% of the total cost. This cost can even be less if the bacteria are only used to start up the biological reactors.

One final technique that can be tested in the treatability study is the addition of organic compounds that can act as co-metabolites. These functions were discussed in the first section of this chapter, and the laboratory treatability study is the first place that these techniques can be tested.

Treatability studies can be done in a variety of hardware. Anything from shaker flasks to respirometers can be used to determine the degradability of the compounds under various situations. A preferred method of the author is the "bag" reactor, Figure 4-28. A felt cloth bag is placed inside of a bucket or barrel. Influent to the reactor goes directly inside the bag. The water travels through the felt and out through the side of the bucket. The water level in the reactor is controlled by the height of the effluent line in the bucket.

Oxygen and mixing are provided by placing an air source at the bottom of the bag. Seed sources of bacteria can easily be placed in the bag. The felt cloth retains the bacteria inside the reactor. Bacteria grow on the felt and in the reactor itself. As can be seen, the reactor is easy to run and can be set up anywhere. The felt does become clogged

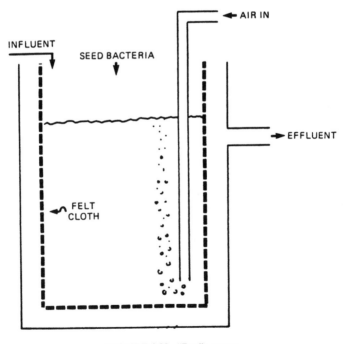

FIGURE 4-28. "Bag" reactor.

over long periods of time, and so the reactor is not a good design for a full-scale plant.

When possible, a pilot plant should be used to confirm the design of a groundwater treatment system. The problem is that many groundwater clean ups rely on relatively small flows. Chapter 6 gives an example in which this problem was overcome by using part of the full-scale system as the pilot plant. This way, the pilot plant confirmed the applicability of the biological treatment. At the same time, all of the parts in the pilot plant were used on the full-scale plant, minimizing the costs for the total project. The biological treatment example in Chapter 6 covers this design example in detail.

Biological treatment will often be used in conjunction with other treatment methods. For strict control of specific organics, it is advisable to use biological treatment followed by carbon adsorption or other technologies. The biological system removes most of the organic compounds at lower cost, and the carbon system ensures that no compounds escape the treatment system. In these cases, the exact concentrations from the biological treatment system are not critical and a pilot plant may not be necessary. Once again, this is especially true when designing small flow systems.

Before we leave this section, it should be noted that a biological treatment system will also strip some organics from the groundwater and soils. The air-water contact for oxygen transfer will also provide air-water contact for the stripping of organic compounds with relatively low vapor pressures. The mixing of the soils will expose surface area for oxygen transfer and for organic transfer to the atmosphere.

It has been the experience of the author and others[47,48] that the rate of degradation is faster than the stripping rate. If the bacteria can degrade a compound, then the driving force behind stripping the compound is the equilibrium concentration of the organic in the reactor. Compounds that are not degradable will have a high equilibrium concentration. If they have a low vapor pressure as well, they will be stripped from the reactor. Compounds that are degradable will have a low equilibrium concentration, and the driving force for stripping will be minimal. The low concentration reactor design uses very low air flow rates and has been shown[49] to strip less than 2% of the organics from treated groundwater at a gasoline station cleanup.

The design engineer should be careful to distinguish between the two removal mechanisms in the design of the treatment system. In situations where there are no regulations covering the discharge of organics to the air, this dual function of the biological system can be an advantage. However, the necessary design additions must be used when the air discharges are regulated. These were covered in Chapter 3, under Air Stripping.

IN SITU TREATMENT

In situ treatment is the destruction, neutralization, and in general, rendering harmless the contaminants while not moving them. The contaminants are treated in place. Theoretically, in situ treatment can be applied to inorganic and organic contaminants. However, in situ treatment is practically limited to the biological destruction of organics. Therefore, in situ treatment will be covered in this chapter even though treatment of inorganics will be discussed.

When the contamination is high pH, low pH, or heavy metals, the main treatment method is to first change the pH (see Chapter 5). In theory, the acid or base could be placed below ground in the path of the groundwater. The groundwater would change its pH as it comes in contact with the neutralizing agent. However, groundwater flows through an aquifer in a plug flow manner. There is very little mixing in an aquifer. The neutralizing agent would be pushed ahead of the groundwater flow, and only come into contact with the leading edge of the plume. In reality, the groundwater must be brought to the surface and neutralized above ground. The treated water can then be returned to the aquifer to force the plume back to the central well. Treatment takes place above ground, not in situ.

Theoretically another method would be to place an immobilized neutralizing agent underground. A slurry wall, made up of limestone, could be placed in the path of the contamination plume. This may work for simple adjustment of an acid plume. The cost, however, would be very high. Also, in the case where heavy metals were present in the water, the precipitation of the metals would clog the slurry wall. In addition, not many regulators would be satisfied with leaving the heavy metals below ground.

The one case where inorganic material can be treated in situ is in the unsaturated zone. The ground can be economically treated in place to a depth where heavy equipment can blend it. The neutralizing agent is placed on the contamination site, and the heavy equipment, normal farming equipment, is used to mix the agent into the soil. Once again, heavy metals are a problem.

Chemical treatment of organic material has the same problems when applied as an in situ technique. Contact and mixing limit the effectiveness of the reaction. The only technology that has been advanced as a viable in situ technique is in situ vitrification. This technology uses electrical energy to turn the sand in dirt into a glass monolith.

The only time in situ techniques provide an added advantage to a treatment technology is with the application of biological treatment. In situ biological treatment has been well documented.[50] The main advantages of in situ biological treatment are:

1. Cost effectiveness,
2. Minimal disturbance to the existing site,
3. On site destruction of contaminants,
4. Continued treatment after shutdown of the project, and
5. Permanent solution.

Applying Biological In Situ Treatment

When applying in situ biological treatment, it is important for the design engineer not to limit his thinking to below ground considerations. Application of in situ techniques require above ground operations as well as below ground operations. All of the bacterial requirements discussed in the biological treatment section of this chapter have to be applied to in situ treatment. The design engineer must satisfy four main requirements:

1. Bacteria,
2. Oxygen,
3. Nutrients, and
4. Environment.

Bacteria do all of the work in biological treatment. The design engineer must make sure that there are sufficient bacteria to consume all of the organic contaminants in a timely manner. The longer the project takes, the higher the cost of the cleanup. The design engineer will have the choice of: enhancing the growth of the existing soil bacteria, growing large amounts of indigenous or imported bacteria in above ground reactors, or applying commercially available bacteria. Once again, it has been the experience of the author that most sites already have the necessary bacteria and that the main problem is removing the limiting factors in the bacterial growth.

When indigenous bacteria are not present, the appropriate bacteria must be placed in contact with the organics. Bacteria are not highly mobile. The bacteria will consume the organics in their immediate vicinity and produce water, carbon dioxide, and new bacteria. Once the food is gone, the bacteria metabolism will slow down and may eventually stop. Bacteria cannot pack-up and move to the next source of food. The design engineer must get the bacteria to all of the organics in an in situ cleanup. Adding bacteria is relatively simple. However, the only way to move any material in an aquifer is by groundwater flow. This is an inefficient method to spread bacteria over large areas.

Excess bacteria may also be produced. When there are large amounts of organics present, the excess bacteria must be removed. Wells and aquifers are made up of small openings. To keep these openings free for water flow, a limited amount of bacteria must be allowed to fill these open spaces. It must be remembered that bacteria only produce between 10 to 20 lbs. of bacteria per 100 lbs. of organic consumed. Therefore, in most situations, the bacteria will take up less space than the original organics. High concentration of starches and sugars may cause the bacteria to produce polysaccharides which can clog small spaces. But, other than these compounds, only inorganic compounds (mainly iron) can clog the aquifer. Screens and wells can still be clogged by a buildup of bacteria as the organics pass through. Clogging is a real possibility and must be part of the design consideration.

The bacteria require oxygen. Aerobic bacteria will degrade organic compounds at a faster rate and leave a lower concentration of specific organics. In addition, aerobic bacteria will yield more new bacteria per pound of organic material consumed. The faster bacteria are produced, the lower the amount of time required for start-up. Some recent work has shown that anaerobic bacteria may be able to treat chlorinated hydrocarbons better than aerobic bacteria.[51] However, in most cases, in situ treatment will be aerobic.

The bacteria require nutrients. The design engineer does not want the level of nutrients to limit the rate of growth of the bacteria. Biological treatment should always be designed to have the organics as the growth limiting factor. Most aquifer contaminations will be produced by the release of pure organic material. It is very unlikely that sufficient nutrients will be available for the bacteria in the spilled material, or in the unsaturated zone and the aquifer. From my experience, the two cases in which ammonia and phosphorous were available occurred when a municipal sewer line leak and a septic tank contributed to the plume. In fact, the presence of ammonia and phosphorous is a strong indication that sewer lines or septic tanks are present. Nutrients will have to be provided for the bacteria. The design engineer must get the nutrients to the bacteria.

One of the problems with nutrient addition to aquifers is the correct form of nitrogen to employ. In above ground biological systems, the nitrogen should be added in a reduced form i.e., ammonia or urea. The bacteria can directly use this form of nitrogen. The author has continued to use the reduced form of nitrogen with in situ programs. Other designers have started to use the oxidized form of nitrogen, i.e, nitrate, as the nitrogen source. No studies could be found which compared the effect of the form of nitrogen on the biochemical reactions in an in situ program. It is suggested that the reader get supporting studies before using the oxidized form of nitrogen on a bioremediation project.

All of the environmental conditions must be appropriate for bacterial growth. Once again, the environmental conditions should not limit the growth rate of the bacteria. The organic materials should always be the growth limit factor. In situ treatment requires that there are no toxic conditions present in the soil or in the aquifer. The pH should be in the correct range and no toxic organics or inorganics should be present. Temperature must also be maintained in the correct range. This will limit the treatment of surface soils during the winter in certain regions of the country. However, groundwater is insulated from the swings in the air temperature, and biological activity can be maintained throughout the year in the aquifer.

Soil conditions provide advantages for biological growth. An advantage of in situ treatment in the unsaturated zone is that the bacteria can tolerate much higher concentrations of toxic organics and inorganics. This is due to the low water content of the soil. Bacteria are only affected by compounds that are in water. Most of the compounds in the unsaturated zone are attached to the soil particles. On the other hand, bacteria are able to degrade the compounds that are not directly in the water. Landfarming of oils is a prime example of this ability. While the concentration of phenol may be 10,000 mg/l in the soil, the bacteria do not see that high a concentration. All of the phenol is not in the water part of the soil. However, when toxic materials are present in the spill area, a treatability study should be run directly on the soil.

Design

The design engineer must take all of the above criteria into consideration when designing an in situ treatment cleanup. We will divide the actual design into three sections:

1. The aquifer,
2. The unsaturated zone, and
3. Above ground equipment.

The Aquifer

The design engineer must control the water movement in the aquifer to ensure the control of the contamination plume, and supply a source of oxygen, nutrients, and bacteria. In Chapter 1 we discussed the plume and the ways to control the movement of the plume by drawing water from central wells and placing the treated water in wells at the outside of the plume. For in situ treatment to work, we must not only control the movement of the plume, we must also get bacteria, oxygen and nutrients spread throughout the affected area of the aquifer. With above ground systems, we had the luxury of using tanks to centralize and control the biochemical reactions. With in situ situations the biological reactor is the entire area of contamination. The design engineer must create a mechanism to deliver the necessary oxygen, nutrients, and bacteria to the reactor.

The first design parameter to determine is the flow. In above ground treatment, this would be the flow to the treatment system. There are two main differences between flow for an above ground treatment system and in situ treatment. First, the flow through the aquifer is the main transport mechanism for the oxygen and nutrients. Since the oxygen and nutrients are usually the limiting factors controlling the cleanup speed, the design engineer normally tries to maximize the rate of flow.

There are five main methods to supply bacteria with oxygen:

1. Injection of aerated water,
2. Air sparging,

3. Injection of hydrogen peroxide,
4. Injection of nitrate, and
5. Venting.

The simplest method of supplying oxygen to the bacteria in the aquifer is to saturate the water with oxygen before injecting it into the ground. This can be accomplished by standard aeration methods or by aerating the water with pure oxygen. Oxygen has a limited solubility, 7 to 10 mg/l, when standard aeration is used. The use of pure oxygen will increase the maximum solubility by a factor of about three. The temperature will affect the solubility in either situation. The water can only supply a limited amount of oxygen. Therefore, highly oxygenated water is recirculated through the aquifer as fast as possible. It is also possible to add oxygen directly to the aquifer by placing intermediate wells between the central wells and the recharge wells, and bubbling air into the standing water of the well.[52,53] Groundwater, passing through the intermediate wells, is supplied with oxygen.

Hydrogen peroxide can also be used as an oxygen source. The hydrogen peroxide decomposes to form water and oxygen. Hydrogen peroxide is very soluble and high concentrations (100 to 1000 mg/l) can be added to the water being injected into the aquifer. However, hydrogen peroxide is also a strong oxidant, and can be harmful to the bacteria if too high a concentration is used.

Recently, nitrate has also been tested as a possible source of oxygen (final electron acceptor) for the bacteria. Once again, nitrate is very soluble in water and high concentrations can be supplied with the injection water. There are two main problems with nitrate. First, all bacteria cannot use nitrate as the final electron acceptor and not all compounds will degrade under nitrate conditions. Second, nitrate is considered a contaminant and the design may have to include the removal of the nitrate at the end of the cleanup.

Finally, oxygen can be supplied by venting the unsaturated zone and allowing the natural transfer of oxygen from the unsaturated zone to the water of the aquifer. This is the only oxygen transfer method that does not relate to the flow of water through the contamination zone. Because of the limited surface area of the water, the transfer is a slow process. But, when venting is used as part of the project, it can be an important aspect of the oxygen supply. The best way to know if venting will assist with oxygen transfer is to analyze the oxygen

TABLE 4-8 Cost of Oxygen Supply for In Situ Treatment

Oxygen-Supply Method	System Flow Rate	Estimated Treatment Time (Days)	Percent of Site Treated	Cost Capital ($)	Operation and Maintenance Cost ($/Month)	Total Cost ($)	Contaminant Treatment Cost ($/lb)
Air sparging	15 wells @ 2 cfm/well	1,716	41	35,000	2,000	148,000	90.3
Water injection	70 gpm	1,580	85	77,000	2,200	191,000	100.2
Venting (vapor control)	160 cfm	132	72	88,500	2,500	99,000	13.4
Hydrogen peroxide	70 gpm	330	95	60,000	11,500	185,000	65.1
Nitrate injection	70 gpm	335	85	120,000	7,500	203,000	77.2

Source: Adapted from Groundwater Technology, Inc. *Hazardous Waste Consultant* (July/August 1990).

202

content of the gases in the vadose zone. Atmospheric oxygen content is about 21%. If the vadose zone gases are significantly below that level, then fresh air brought into the unsaturated zone will increase the rate of oxygen supply. The author has worked on projects where the oxygen content of the unsaturated zone was less than 2%. Venting was successfully applied at that location.

All of these methods have advantages and disadvantages. The design engineer will have to pick the right method for each particular site. Table 4-8 shows a cost comparison from the literature[54] of the various methods of oxygen supply. These numbers should only be used as background information, and cost analysis should be completed for each unique site.

The flow will also affect the supply of nutrients to the contamination area. Nutrient addition in the form of ammonia and phosphate are not limited by solubility. However, it is not a good idea to place too much nutrients in the aquifer at one time. Both organic contaminant and nutrient should be used up at the end of an in situ cleanup of an aquifer.

Part of the operations of an in situ cleanup must be the testing, for residual nutrients, of the water drawn from the aquifer. A minimal amount of nutrients should be applied when treating the aquifer. The first section of this chapter provides the details on nutrient requirements of the bacteria. The soil of the aquifer can affect the nutrients by adsorption or reaction. Tests should be completed to discover any interaction between the aquifer sediments and the nutrients. Sometimes the form of the nutrient can be changed so that it no longer interacts with the soil but is still available for the bacteria.

There are several reasons for this. High concentrations of ammonia can change the pH or may be directly toxic to the bacteria. There is also the chance of secondary plumes of nutrients. Finally, excess ammonia can create it's own oxygen demand and create nitrates.

The second difference with water flow in an in situ treatment is that the water being recharged to the aquifer is not completely clean. The water entering the aquifer must have oxygen and nutrients. The presence of adapted bacteria in the recharge water can also help to speed the cleanup. Figure 4-29 shows the zone of influence of the draw down by the central well. The recharge water should not be placed outside of this zone of influence. The bacteria and, more likely, the nutrients could be considered a contaminant downflow in the aquifer.

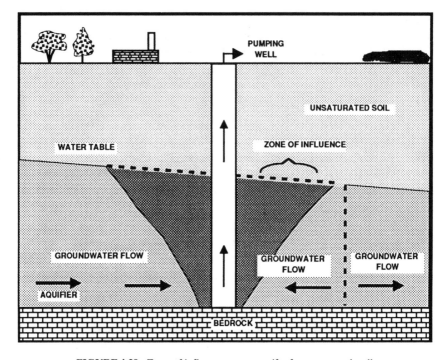

FIGURE 4-29. Zone of influence on an aquifer from a central well.

The only way to ensure that the water placed in a recharge well does not get out of the zone of influence is to place less water in the recharge well than was drawn from the central well. The zone of influence extends all around the central well. The plume travels in one direction from the well. The recharge well is placed at the end of the plume. While the recharge water will reduce the amount of water taken from the opposite flow side of the central well, there will still be some water from the upflow side. The design engineer will have to work closely with the hydrogeologist to determine the correct ratio of the two flows. The excess water will have to be discharged off site or used as part of the cleanup of the unsaturated zone.

The Unsaturated Zone

Most contaminations of aquifers are a result of material being released above the aquifer in the unsaturated zone. If the cleanup is limited to the aquifer, the contaminants still in the unsaturated zone can be a source of future contamination. In situ cleanup techniques can also be applied to the unsaturated zone. The design criteria is the same as that for in situ cleanup of an aquifer, with two exceptions. First, the unsaturated zone may not require any water pumping as part of the cleanup. Second, the objective in the unsaturated zone may be flushing of the contaminants along with in-place destruction.

Treatment in the unsaturated zone has the same bacterial requirements as does treatment in the aquifer. There must be sufficient bacteria, oxygen, nutrients, and the correct environmental conditions. Temperature problems in the winter may prevent in situ cleanup during winter months in some regions of the country. One difference in the treatment criteria is that there must also be sufficient water present in the unsaturated zone. When the cleanup of an unsaturated zone is part of an aquifer cleanup, there is always sufficient water available. In cases where only the unsaturated zone is being cleaned, water may have to be supplied. This is especially true in the arid regions of the country.

The upper three to seven feet, depending upon the type of soil, can be reached by normal farming implements. Therefore, these upper layers of the unsaturated zone can be treated by landfarming techniques. Landfarming is the in-place destruction of organic material with bacteria. These techniques have been well established for the treatment of oily waste from the petroleum industry. These same techniques have now been used on land contaminated with hazardous waste. A prime example is the cleanup of wood treating sites by landfarming. As discussed in soil reactor sections, bacteria have been successful in removing pentachlorophenol and a range of other phenol based compounds. The one problem with landfarming is the regulatory control on this process. Regulators have taken a dim view on the classic landfarming technique of spreading contaminants on clean soil in order to "farm" it. Any landfarming proposal will come under close scrutiny. However, usually as the land is already contaminated, a design that includes the supply of nutrients and oxygen to speed a biological destruction (landfarming) will be allowed.

There are two basic ways in which to enhance biological activity in the unsaturated zone; water movement through the zone or air movement through the zone. As we have discussed several times, the bacteria need oxygen and nutrients in order to increase their rate of reaction and to clean up the contaminated area. The main function of either method is to deliver the required compounds.

Water can be introduced at the surface, in recharge trenches or in shallow recharge wells. The important part of the design is to ensure that the water travels throughout the entire contaminated zone. Oxygen and nutrients can be added to the water as discussed in the previous section. Bacteria can be added if required, but care must be maintained so that the bacteria do not clog the recharge trench or well.

A difficult design factor for water movement through the unsaturated zone is the collection of the water after it moves through the contaminated zone. The water will probably solubilize small amounts of the contaminant as it comes in contact with the zone. The water can be collected before it comes into contact with the aquifer, or the water can be allowed to enter the aquifer. The problems with collecting the water in an unsaturated zone are many, and the designer should be very careful if this method is chosen. The aquifer should only be used as the collection method when it is also contaminated.

Air movement through the unsaturated zone can also be used to enhance biological activity. Oxygen in the air can be used as a direct source of oxygen for the bacteria in the unsaturated zone. Nutrients and bacteria cannot be delivered by this method. However, oxygen alone can greatly enhance the rate of biological activity. The main design criteria is to place collection and injection (if required) locations to ensure that the air moves through the entire contaminated zone.

Above Ground Equipment

All in situ cleanups will require above ground equipment. Landfarming needs farming equipment to spread the bacteria and nutrients, and to provide mixing and oxygen. Aquifer cleanups will require a minimum of pumps and mix tanks for the supply of bacteria, nutrients, and oxygen. In addition, most aquifer and unsaturated zone cleanups will require a treatment tank above ground. This treatment tank will be used for four main functions:

1. Produce bacteria,
2. Reduce organic content of the water,
3. Addition of oxygen to the water, and
4. Addition of nutrients to the water.

Bacteria are the work horse of an in situ cleanup. Large quantities of bacteria are required for the process. Natural bacteria are usually sufficient for the most cleanups. However, an above ground biological unit can increase the rate of cleanup. A standard biological treatment system can be set up above ground. The system can be a lagoon, activated sludge, or fixed film system, depending on the amount of bacteria required.

Bacteria will produce at a rate of between 0.10 and 0.25 pound of bacteria per pound of organic material consumed. Any of the biological treatment systems discussed in the first section of this chapter are very effective in growing the required bacteria. The choice among the systems will be made mainly on the amount of bacteria required and the concentration of contamination allowed to be recharged to the aquifer. Lagoons produce the highest quantity of bacteria per pound of organic material. However, lagoons produce too high a concentration of bacteria when a well recharge is to be used. The bacteria can clog the well. Activated sludge or fixed film systems produce the highest quality organic efficient content. Activated sludge or fixed film systems with a clarifier will both discharge lower concentrations of bacteria. Settled bacteria in the clarifier can be added to the recharge water to obtain the desired bacterial concentration.

The food for the bacteria is readily available. The central well will be in the center of the plume. The contaminants will be at their highest concentration at this point. The water drawn from the well is sent to the treatment tank, and the bacteria grow on the contaminants. This will produce bacteria that are actively degrading the compounds found in the aquifer. The same holds true for compounds flushed from the unsaturated zone.

The reaction tank accomplishes the second function simultaneously with the first function. The bacteria use the contaminants found in the groundwater for food. As they use the compounds, the organics are removed from the water. The recharge water has the bacteria in it, but not the original contaminants. The recharge water should not have a high concentration of the contaminants in it.

The final uses of the above ground reaction tank are to supply nutrients and oxygen. The nutrients are added to the tank. Enough nutrients must be added to satisfy the requirements of the biological reaction within the tank, and to have a residual for the recharge water. The same is true for oxygen. The discharge concentration of oxygen from the reaction tank should be near the saturation level. This is true even if hydrogen peroxide or nitrate is used as an oxygen source. While in situ treatment will lower the cost of the project and speed its completion, the rate of biological destruction is always faster in an above ground tank because oxygen and nutrients can be readily supplied. Any contaminants that can be destroyed above ground, should be. Also, the bacteria will be "hungry" for more food if they are reintroduced to the ground. Removal of the contaminants will also minimize the oxygen requirements below ground.

NATURAL BIOCHEMICAL REACTIONS

One final area needs to be covered before we leave this chapter, natural biodegradation. Aquifers and vadose zones are living environments. Natural bacteria will have an effect on the contaminants. These effects must be incorporated in the investigation, design, and long-term operation, and monitoring of a site.

A recent evaluation[55] of actual concentrations found in groundwater and expected concentrations based on retardation indicate that a solute transport model is likely to be overly conservative in its estimation of the transport of some compounds, particularly the compounds subject to biodegradation. Figure 4-30 shows the extent of the total volatile organic plume at the site. Figures 4-31 through 4-34 show the theoretical and actual distribution for four specific compounds.

Except for 1,4-dioxane, Figure 4-31, which has traveled almost as far as expected, the other compounds have migrated smaller distances than had been anticipated. Tetrahydrofuran, Figure 4-32 and dioxane,[1,4] which have identical retardation factors and similar solubilities in water exhibit greatly different migration rates; 1,4-dioxane has traveled about 2.5 times further than tetrahydrofuran. Tetrahydrofuran is amenable to biodegradation (although at a slow rate), whereas 1,4-dioxane is not. In addition, benzene, Figure 4-33, and phenol, Figure 4-34, which are also amenable to biodegradation should have traveled 1,250 and 2,500 feet, respectively, beyond the site boundary. The analytical data indicate that these two compounds have migrated less than 100 feet from the site boundary. They are readily degraded.

The actual degradation rate of these four compounds depends on many factors. Are there bacteria present that recognize the compounds as a food and energy source? Are there cometabolites present

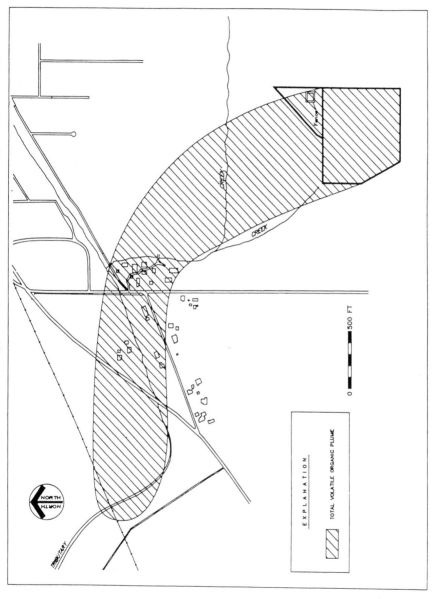

FIGURE 4-30. Extent of total volatile organic plume.

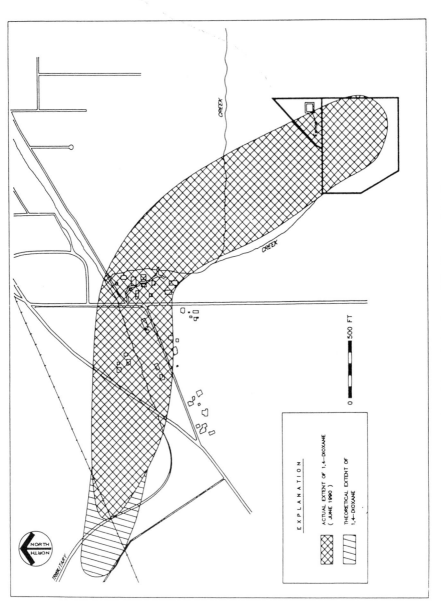

FIGURE 4-31. Theoretical and actual distributions of 1,4-dioxane.

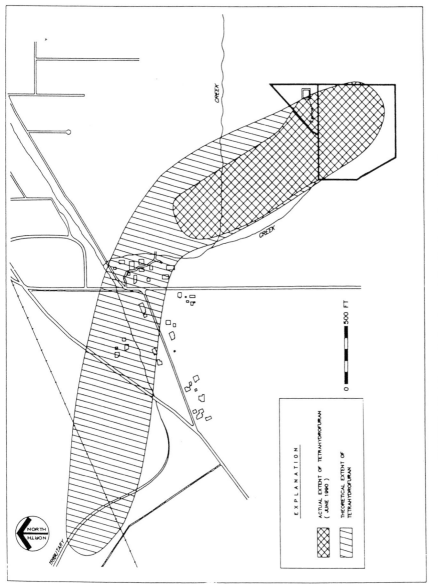

FIGURE 4-32. Theoretical and actual distributions of tetrahydrofuran.

211

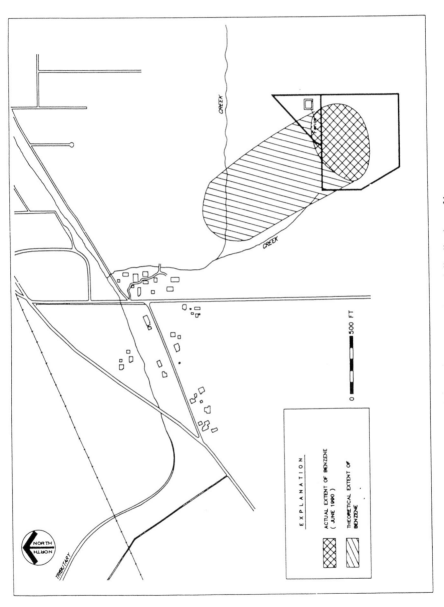

FIGURE 4-33. Theoretical and actual distributions of benzene.

212

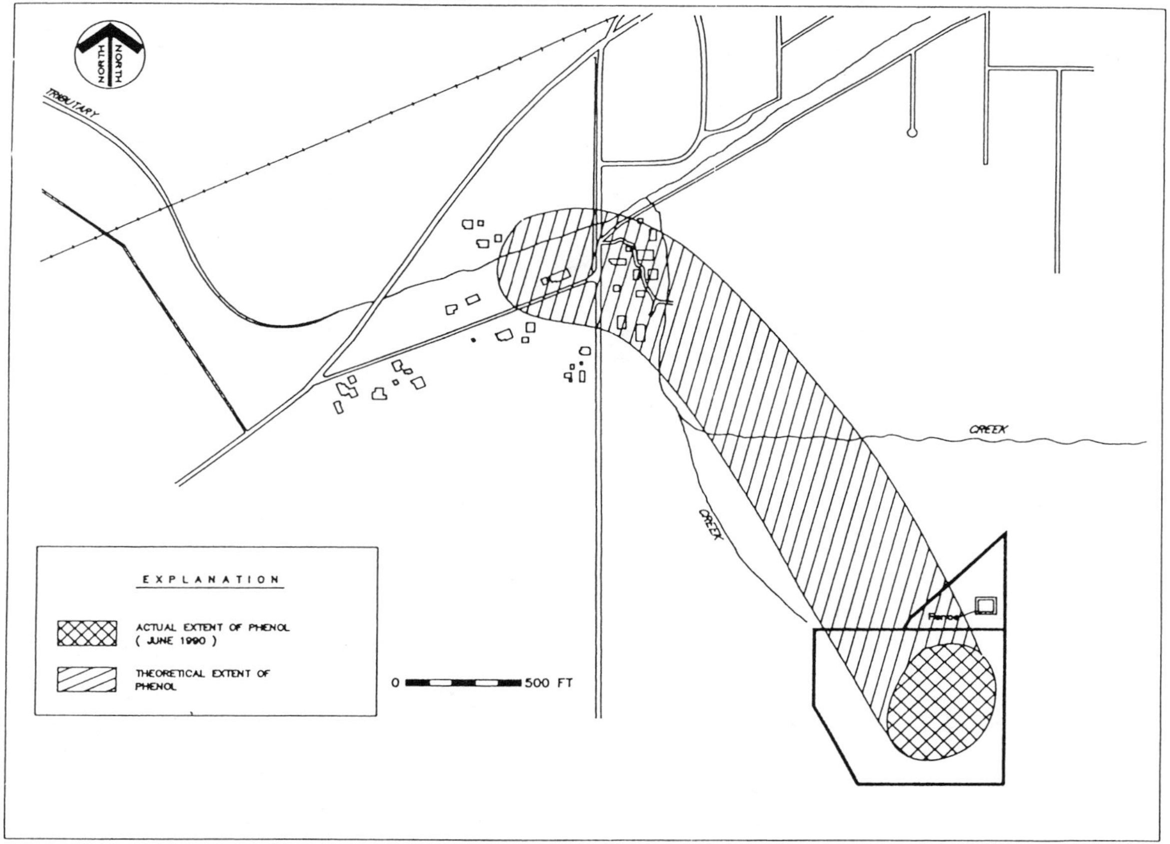

FIGURE 4-34. Theoretical and actual distributions of phenol.

213

214 Groundwater Treatment Technology

for the nondegradable compounds? Are the environmental conditions optimum for bacterial activity? Are there nutrients present? How much oxygen is naturally present for aerobic degradation? This study showed that the compounds that will only degrade under aerobic conditions, THF, benzene and phenol, only start to disappear after the plume has exited the area covered by the interim cap. Tests have shown that small amounts of oxygen also start to appear in the area after the cap.

Summary

Biological treatment is an important technology for remediation of contaminated sites. This chapter has provided a detailed review of the bacteria, their biochemical reactions, the reactors in which they grow, and the requirements for in situ treatment.

The most important points to remember about biological treatment is that it is a natural process and that it is probably already occurring at the site. The main objective of biological remediation designs is to remove the limiting factors in the growth of the bacteria. We are not bringing a new process to the site, we are enhancing a natural process that is already occurring at the site.

References
1. Superfund: The stuff myths are made of. *Civil Engineering News* January 1990.
2. *Bergy's Manual of Determinative Bacteriology,* 8th ed. Williams & Wilkins, Baltimore, Maryland.
3. Paul, E. A. and Clark, F. E. In *Soil Microbiology and Biochemistry,* Academic Press, San Diego, 1989.
4. Bouwer, H. In *Groundwater Hydrology.* McGraw-Hill, New York, 1978.
5. Dragun, J. Microbial Degradation of Petroleum Products in Soil; In *Proceedings of a Conference on Environmental and Public Health Effects of Soils Contaminated with Petroleum Products,* October 30-31, 1985, University of Massachusetts, New York, John Wiley & Sons, 1988.
6. Kobayashi, H. and Rittmann, B. E. Microbial removal of hazardous organic compounds. *ES&T* 16:170A-183A, 1982.
7. Bumpus, J. A., Tien, M., Wright, D., and Aust, S. D. Oxidation of persistant environmental pollutants by a white rot fungus. *Science* 228:1,434-1,436, June, 1985.
8. Brock, T. D. In *Biology of Microorganisms.* Prentice-Hall, Englewood Cliffs, New Jersey, 1979.

9. Freeman, H. M., Ed. In *Standard Handbook of Hazardous Waste Treatment and Disposal*. McGraw-Hill, New York, 1989.

10. Stanier, R. Y., Adelberg, E. A., and Ingrahm, J. L. In *The Microbial World*. Prentice-Hall, Englewood Cliffs, New Jersey, 1976.

11. Viessman, W. and Hammer, M. J. In *Water Supply and Pollution Control*, 4th Ed., Harper & Row, New York, 1985.

12. Moat, A. G. *Microbial Physiology*. Wiley-Interscience, New York 1979.

13. Dragun, J. In *The Chemistry of Hazardous Materials*. The Hazardous Materials Control Research Institute, Silver Springs, Maryland, 1988.

14. Grady, C. P. Biodegradation: it's measurement and microbial basis. *Biotechnology and Bioengineering* 27:660-674, 1985.

15. Bull, A. T. *Contemporary Microbial Ecology*. D. C. Ellwood, J. N. Hedger, M. J. Lathane, J. M. Lynch and J. H. Slater Eds., Academic Press, London, 1980.

16. Alexander, M. Biodegradation of organic chemicals, *ES&T* 19(2), 1985.

17. Atlas, R. M., Ed., *Petroleum Microbiology*. Macmillan, New York, 1984.

18. Brink, R. H., Biodegradation of organic chemicals in the environment. In McKinney, J. D. (ed), *Environmental Health Chemistry*. Ann Arbor, MI.

19. Vogel, T. M. and McCarty, P. L., Transformations of halogenated aliphatic compounds. *ES&T* 21(8), 1987.

20. Ghosal, D., You, I. S., Chatterice, D. K., and Chakrabarty, A. M. *Science* 228(4696), 1985.

21. Rittman, B. E., Smets, B. F., and Stahl, D. A. The role of genes in biological processes Part V. *ES&T* 24(1):23-29, 1990.

22. McCarty, P. L. Anaerobic biotransformation of chlorinated solvents. Abstract in *Biological Approaches to Aquifer Restoration, Recent Advances and New Opportunities*. Standford University, 1986.

23. Vogel, T. M. and McCarty, P. L. Abiotic and biotic transformations of 1,1,1-trichloroethane under methanogenic conditions, *ES&T* 12:1208-1213, 1987.

24. Kleopfer, R. D. et al. Anaerobic degradation of trichloroethylene in soil, *ES&T* 19:277-280, 1985.

25. Parsons, F., Wood, P. R., and DeMarco, J. Transformations of tetrachloroethane and trichloroethane in microcosms and groundwater, *AWWA Journal* 76:56-59, 1984.

26. Barrio-Lage, G., Parsons, F. Z., Nassar, R. J., and Lorenzo, P. A. Sequential dehalogenation of chlorinated ethenes. *Environ. Science Technol.* 20:96-99, 1986.

27. Cooper, W. J. et al. Abiotic transformations of halogenated organics, l. Elimination reaction of 1,1,2,2-tetrachloroethane and formation of chloroethene. *ES&T* 21:1112-1114, (1987).

28. Wood, P. R., Lang, R. F., and Payan, I. L. In *Groundwater Quality,* Ward, C. H., Giger, I. S., and McCarty, P. L., Eds. Wiley & Sons: New York, 1985.

29. Nyer, E. K. and Ziegler, D. Hazardous waste destruction by submerged fixed-film biological treatment. *Fifteenth Mid-Atlantic Industrial Waste Conference.* 1983.

30. Hildebrandt, W. W. and Wilson, S. B. On-site remediation of organically impacted soils on oilfield properties. Given at the California Regional Meeting of the Society of Petroleum Engineers, Ventura, California, April 1990.

31. Barnhart, M. J. and Myers, J. M. Pilot bioremediation tells all about petroleum contaminated soil. *Pollution Engineering,* 21(10):110, (1989).

32. Hutzler, N. J., Baillod, C. R., and Schaepe, P. A. Biological reclamation of soils contaminated with pentachlorophenol. In *Proceedings of the Sixth National Conference on Hazardous Wastes and Hazardous Materials.* New Orleans, Louisiana, p. 361, April 1989.

33. Loehr, R. C. and Kabrick, R. M. Bioremediation of contaminated soils. In *Proceedings of the Sixth National Conference on Hazardous Wastes and Hazardous Materials.* New Orleans, Louisiana, p. 301, April 1989.

34. Linkenheil, R. J. and Patnode, T. J. Bioremediation of contamination by heavy organics at a wood preserving plant site. In *Superfund '87: Proceedings of the 8th National Conference.* Washington, D.C., p. 193, November 1987.

35. Lynch, J. and Genes, B. R. Land treatment of hydrocarbon contaminated soils. In *Petroleum Contaminated Soils,* Volume 1, P. T. Kostecki and E. J. Calabrese (Eds.), Lewis Publishers, Inc., Chelsea, Michigan, p. 175, 1988.

36. Appendix C: Biological treatment technologies. *Technology Screening Guide for Treatment of CERCLA Soils and Sludges.* EPA/540/2-88/004, p. 103, September 1988.

37. Williams, R. T. and Ziegenfuss, R. S. Composting of explosives and propellant contaminated sediments. In *Proceedings of the Third International Conference on New Frontiers for Hazardous Waste Management.* Pittsburgh, Pennsylvania, p. 204, September 1989.

38. Bourquin, A. W. Bioremediation of hazardous waste. *Hazardous Materials Control* 2(5):16 (1989).

39. Golueke, C. G. and Diaz, L. F. Biological treatment for hazardous wastes. *BioCycle,* 30(12):58 (1989).

40. Stroo, H. F. Biological treatment of petroleum sludges in liquid/solids contact reactors. *Environmental and Waste Management World* 3(9):10 (1989).

41. U.S. Environmental Protection Agency. *Treatment Technology Bulletin: Slurry Biodegradation.* To be published by the U.S. Environmental Protection Agency in May 1990.
42. Brox, Gunter H. and Hanify, D. E. A new solid/liquid contact bioslurry reactor making bio-remediation more cost-competitive. Given at the Colorado Hazardous Waste Management Society Conference, Denver, Colorado, November 1989.
43. Telephone interview with Gunter H. Brox of EIMCO Process Equipment Company, Salt Lake City, Utah, April 12, 1990.
44. Torpy, M. F., Stroo, H. F., and Brubaker, G. Biological treatment of hazardous waste. *Pollution Engineering* 21(5):80 (1989).
45. Bhattacharyya, P. E., Fu, J., and Smith, J. R. Biological treatment of contaminated soil on closed coke coal tar distillation and gas manufacturing plants. Given at The 1988 AISI Convention, Cleveland, Ohio, September 1989.
46. Kroos, H. Regional biological decontamination centers for the clean-up of contaminated soil, sludges and industrial waste-waters, In *Proceedings of the Forum on Innovative Treatment Technologies: Domestic and International.* Atlanta, Georgia, p. 124, June 1989.
47. Kincannon, D. F. et al. Predicting treatability of multiple organic priority pollutant wastewaters from single pollutant treatability studies. 37th Purdue Industrial Waste Conference, May 1982.
48. Freeman, R. A. et al. Air stripping of acrylonitrile from wastewater systems. *Environmental Progress* 3:1-26 (1984).
49. Nyer, E. K. et al. Decay theory biological treatment for low level organic contaminated groundwater. *Proceeding of the Second National Outdoor Conference on Aquifer Restoration.* May 1988.
50. Lee, M. D. and Ward, C. H. *Biological Methods for the Restoration of Contaminated Aquifers.* National Center for Groundwater Research, 1983.
51. Bouwer, E. J. and McCarty, P. L. Modeling of trace organics biotransformation in the subsurface. *Ground Water* 22:4-433 (1984).
52. Yaniga, P. M. Groundwater abatement techniques for removal of refined hydrocarbons. *Hazardous Wastes and Environmental Emergencies Proceedings.* March, 1984.
53. Jhaveri, V. *Bio-Reclamation of Groundwater By The GDS Process.* Groundwater Decontamination Systems, Inc.
54. Methods of supplying oxygen for in situ bioremediation. *The Hazardous Waste Consultant* July/August 1990.
55. Nyer, E. K. et al. Biochemical effects on contaminated fate and transport. *Groundwater Monitoring Review* Spring 1991.

5

Treatment Methods for Inorganic Compounds

The main inorganic contaminants found in groundwater include:

Heavy metals,
   Arsenic,
   Cadmium,
   Trivalent chromium,
   Hexavalent chromium,
   Copper,
   Lead,
   Mercury,
   Nickel,
   Silver,
   Zinc,
Nitrates,
Sulfates,
Total dissolved solids,
High and low pH.

Inorganic contaminants in groundwater have not had as much attention as have the organic contaminants in the past few years. There are several reasons for this. First, the unsaturated ground, vadose zone, has a limited ability to remove these contaminants from a spill, so that they never reach the aquifer. Most soils have an ion exchange capacity. A heavy metal moving through the soil will be exchanged with a cation in the soil and be adsorbed by the soil. In addition, anaerobic zones in the soil can biologically transform nitrates into nitrogen gas. The soil will also have a natural ability to neutralize pH to a limited extent.

Inorganic compounds are also not used as often as organic compounds for industrial purposes. Industrial plants do not typically put heavy metal solutions into large storage tanks and pipelines are not typically used to transfer these compounds. Heavy metals in their pure state are not soluble in water. Even when stored as a salt, which is soluble in water, they are in a solid form. There is no equivalent for inorganic compounds to gasoline storage tanks, oil pipelines, solvent storage tanks, etc.

The largest source of heavy metal contamination is leachate from abandoned waste disposal sites. Legal landfills, illegal landfills, and old mines are probably the major sources of heavy metals. Ten years ago, the regulations did not anticipate the problems caused by improper burial of heavy metals.

The largest source of arsenic, nitrates, and total dissolved solids contamination is agricultural chemical usage. These sources represent low concentration material being introduced into the ground. Chemicals used for agriculture are also a source of organic contaminants, mainly pesticides.

Low concentrations of nitrates have been known for a long time to cause methemoglobinemia in infants. Recently, the public has recognized that other low concentrations of inorganic compounds are also not acceptable because of health concerns. This, combined with the cleanup of abandoned sites, has made the inorganic materials area more important.

This chapter will cover the following methods for removal of inorganics from groundwater:

1. Chemical addition,
2. Removal of suspended solids,
3. Reverse osmosis, electrodialysis, distillation, and
4. Ion exchange.

A major advantage of developing treatment methods for inorganics in groundwater is that most of the methods described in this chapter can be tested in the laboratory. The results can be accurate enough to develop a preliminary design of the full scale system, however, pilot studies should be conducted for any of the methods that use membranes, i.e., reverse osmosis. Pilot studies may also be warranted if there is a "mixed bag" of contaminants requiring removal. The laboratory test for inorganic compounds can also be performed in a short period of time, days, versus the weeks and months that it takes to get good data for organic treatment methods. All of the methods listed in this chapter should be tested at several different concentrations and pH ranges. All water is slightly different, and while dosages and pH levels are recommended in the text, the optimum conditions will have to be found for treating each groundwater situation.

CHEMICAL ADDITION

pH Adjustment

There are two main purposes for pH adjustment in the treatment of groundwater. The first is the adjustment of the groundwater to a neutral pH of around 7. Water that is to be discharged to a receiving stream must have a pH of between 6 and 9. The second is for the precipitation of heavy metals. While precipitation normally includes an adjustment of pH, other chemical addition is necessary when removing chemicals such as arsenic and hexavalent chromium. This will be covered in the next section.

In wastewater treatment, the preferred method of pH adjustment is to merge two waste streams of differing pH. During groundwater treatment, a secondary source for pH adjustment is usually not available. Therefore, the main methods used for adjustment of acidic water are:

1. Passing through a limestone bed,
2. Mixing with lime slurries,
3. Adding caustic soda, NaOH, and
4. Adding soda ash, Na_2CO_3.

The main methods for adjustment of alkaline water are:

1. Bubbling carbon dioxide in the water, and
2. Adding a strong acid, HCl, H_2SO_4, etc.

It is very rare for a groundwater to be too alkaline. The main reason for adding acids to a groundwater is for pH readjustment after the water has been raised to a high pH in order to precipitate a metal. A strong acid addition is the typical method used for this adjustment. The proximity of an industrial plant impacts the cost effectiveness of carbon dioxide addition.

It should be noted that aeration treatment methods for the removal of volatile organic compounds, i.e., air stripping, diffused aeration, etc., will liberate carbon dioxide and slightly raise the pH of groundwater. Generally, the pH adjustment is a unit or less, i.e., from 7 to 8.

Depending on the final pH required, lime can take up to a 30 minute reaction time to be completely used. The slurry should be introduced into the acidic groundwater in a completely mixed tank. The residence time in the tank should therefore be 30 minutes. (Once again, this number does not represent a hard design number. Test the actual reaction time on each groundwater.) This slow reaction will make pH control more difficult. Lime will also form more sludge than would neutralization with caustic.

Caustic can be delivered and stored as a liquid. The reaction time is very fast with caustic. The reaction tank should still be completely mixed, and about five to ten minute residence time is usually sufficient. Caustic can be used at several different concentrations. This can be an advantage for small streams that are near neutral pH. Short term, one to two years, groundwater treatment projects will probably use caustic for pH adjustment. Long term projects may take advantage of the low cost of lime.

Chemical Addition Before Precipitation

The main reason for pH adjustment is to remove heavy metals from the groundwater. The pH will normally have to be raised above 7 to remove the metals. This is especially true when the metals are held in solution by a chelant. Figure 5-1 summarizes the solubilities of the various heavy metals as a hydroxide precipitate. For metals precipitation, lime or caustic (NaOH) are necessary to reach required pH ranges. The main differences between the two compounds are that lime is less expensive but is more difficult to use. Lime costs about 40 to 60% less than caustic. Actual prices will depend heavily on transportation costs. Specific prices should be obtained for each project. Lime is normally fed as a hydrated lime slurry. The material is stored dry and slurried before mixing with the acidic groundwater. Lines can become clogged easily, the great care should be taken to keep the slurry in motion.

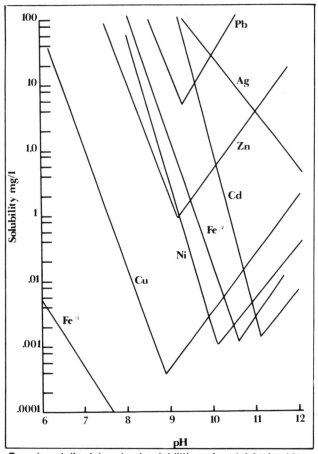

Experimentally determined solubilities of metal hydroxides.

FIGURE 5-1. Solubilities of metal hydroxides at various pH's. (Courtesy of Graver Water.)

Not all metals will precipitate upon an increase in pH. If the metals are being kept in solution by a chelant, the metals will not follow the solubility curves in Figure 5-1. In addition, iron in the ferrous state and chromium in the hexavalent state will not precipitate at high pH. Arsenic is another inorganic compound that will not precipitate by a simple increase in pH. All of these compounds require chemical addition before they can be precipitated.

Another way to remove metals from solution is to precipitate them as a sulfide precipitate as opposed to a hydroxide precipitate. The solubility is still dependent on the pH, but in general, a metal sulfide is

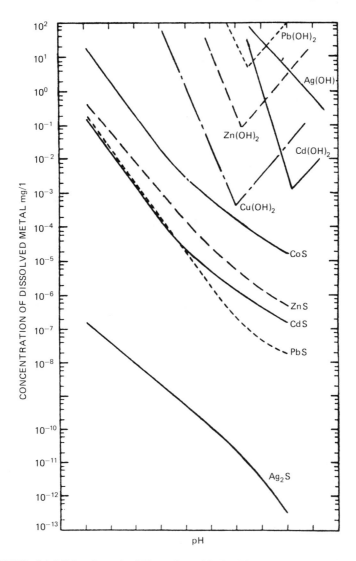

FIGURE 5-2. Solubilities of metal sulfides and metal hydroxides at various pH's. (Courtesy of Graver Water.)

less soluble than a metal hydroxide. Since the solubility is less, they will precipitate out of the water resulting in the water effluent concentration of the metal being less. Also in the case of metals like lead and zinc, the range of pHs that the reaction can be run in is increased. Figure 5-2 summarizes the solubility of metal sulfide compounds. Long term projects may want to consider this technology if there is a reason that very low metal concentrations are required in the effluent from the treatment system.

Iron is not a toxic metal. However, it can cause problems with processes, pipes, equipment, and final use of the water. As presented in the air stripping section of Chapter 3, iron is a nuisance metal that has to be considered when designing a treatment system for VOC removal. Iron is found in many groundwaters naturally. The technology for removing iron is well established. Soluble ferrous iron has to be oxidized to the insoluble ferric state. In the ferric state, iron is not soluble above a pH of 7. Oxidation occurs readily at pH 7.0 to 7.5. The water must simply be aerated within this pH range and ferrous iron will convert to ferric iron. Diffused air aerators are one means of providing oxygen to the water. Amounts of air used vary from 0.04 to 1.5 m^3/m^3 of water.[1] A reaction time of up to 30 minutes should be allowed.

The addition of sodium or calcium hypochlorite could also be utilized to oxidize iron from the ferrous to the ferric state. They are available in liquid form generally within the 12 to 15% concentration range. They degrade with time and therefore cannot be stored for long periods. Generally, no more than one month's storage is practical. The problem with using these chemicals is that they introduce chlorine into the waste stream and have the potential to form more trichloromethanes and other chlorinated compounds. Therefore, their use should only be considered when air emissions are a serious concern and low levels of trichloromethanes or trihalomethanes are acceptable to discharge. Hydrogen peroxide can also be used as a chemical oxidant for iron. It is more expensive than sodium or calcium hypochlorite but it has a longer shelf life and does not cause the problems of creating chlorinated hydrocarbons.

Hexavalent chromium is a toxic heavy metal. Hexavalent chromium is only used in industrial plants and is not found naturally in groundwater. Like ferrous iron, it is soluble in water at high pH. Methods for treating hexavalent chromium have been developed for industrial wastewater. The hexavalent chromium must be reduced to the trivalent state for removal. Trivalent chrome is not soluble at high pH. (See Figure 5-1). The standard reduction treatment technique is to lower the pH of the water to 2.0 to 3.0. Next, a chemical reducing agent is added to the water. The most common reducing agent is sulfur dioxide, but sodium bisulfite, metabisulfite, hydrosulfite, or ferrous sulfate are also used. The pH is then raised and the trivalent chromium precipitates.

Hexavalent chromium reduces readily but not at a fast rate. The longer that the hexavalent chomium is in the groundwater and the more contact it has with other material and water, the more likely it will naturally be reduced to the trivalent state. However, when hexavalent chomium is found in the groundwater to be treated, at least a 20 to 30 minute residence time should be used for the reduction reaction in the treatment system. As with all inorganic treatment methods, the reaction time for the hexavalent chomium reaction should be established by laboratory tests.

Mercury precipitation as an hydroxide is ineffective since mercury hydroxide is soluble over a wide pH range. Sulfide addition, in order to precipitate highly insoluble mercury sulfide, is the most common precipitation treatment reported.[3] Other means of treatment consist of ion exchange, carbon adsorption, and coagulation. It has been reported by Patterson[3] that the lower treatability limit for mercury is 10 to 20 mg/l with sulfide precipitation, 1 to 5 mg/l with ion exchange, 1 to 10 mg/l with alum coagulation, 0.5 to 5 mg/l with iron coagulation, and 0.25 to 20 mg/l with activated carbon depending upon initial concentration.

One final inorganic that is typically removed by chemical addition is arsenic. Arsenic in the groundwater may be either in the arsenite ion (AsO^{2-}) or the arsenate ion (AsO^{4-}) form. Since the presence of dissolved oxygen will oxidize arsenite to arsenate, the majority of groundwater contamination cases involve the arsenate form. If a deep aquifer is contaminated, the groundwater may have to be oxidized prior to arsenic treatment since the arsenate form is more easily removed than is the arsenite form. Arsenic requires the formation of a floc in order to be removed from water. Increasing the pH with lime will remove some of the arsenic. The same increase in pH with caustic soda will not remove the arsenic, because of the lower amount of solids formed. Arsenic treatment processes for both municipal wastes and industrial wastes, insofar as the available literature indicates, are similar and commonly involve coprecipitation by addition of a polyvalent metallic coagulant, to produce a hydroxide floc.[3] It has been the author's experience that the most efficient way to remove arsenic is to add iron, in the ferrous or ferric state, at a pH of between 5 and 6, and then to increase the pH to 8 to 9 with lime. Iron/arsenic ratios

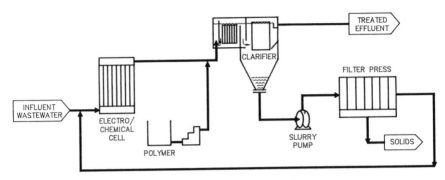

FIGURE 5-3. Process flow diagram electrochemical precipitation. (Courtesy of Andco.)

should be around 8:1. Values above or below that ratio will decrease the removal efficiency.

Before we leave this section, one further method of precipitation should be reviewed. Iron can be added to the water by electrochemical methods to enhance the precipitation of other inorganics. Electrochemical technology for the treatment of heavy metals is a relatively new technology which has been seriously used only since the early 1970s. The system uses sacrificial electrodes to produce an insoluble ferrous ion, which adsorbs and coprecipitates heavy metals. The mechanism is not fully understood, but it is believed to be primarily an adsorption process with the metals being adsorbed onto the iron hydroxide matrix that is formed in the electrochemical cell. The process flow diagram is shown in Figure 5-3.

The electrochemical process can be operated at neutral pH which may eliminate the additional pH adjustment step for metal treatment. A typical electrochemical process unit is presented in Figure 5-4. In the process of producing the ferrous ion, the electrodes are consumed and will require replacement. To keep the electrodes clean, diluted acid wash is used approximately once a day. Most of the maintenance can be set up to function on an automatic basis. There is a large amount of field experience on treating industrial waste, but this technology has only recently been applied to groundwater treatment. However, the ability to design the technology for small flows and to have automatic operation make it a good technology for groundwater treatment.

FIGURE 5-4. Electrochemical treatment system. (Courtesy of Andco.)

REMOVAL OF SUSPENDED SOLIDS

Flocculation

By chemical addition and pH adjustment, the inorganic contaminants are converted to a nonsoluble form. All the waste is now contained in suspended solids that must be removed from the water. These suspended solids are all heavier than water. However, these solids will not be effectively removed simply by placing the groundwater in a quiescent tank and allowing the particles to settle.

For example, assume that all of the particles have a specific gravity of 2.65. The particles that have a diameter of 0.1 mm, about the size of fine sand, will take 38 seconds to settle one foot. Particles with a 0.01 mm diameter, about the size of silt, will take 33 minutes to settle the same foot. Finally, particles with a diameter of 0.001 mm, about the size of bacteria, will take 55 hours to settle one foot. Colloidal particles would take years to settle the same distance.[2]

The sizes of the particles generated by the chemical addition and pH adjustment will range from colloidal to fine sand. The specific gravity will be lower than 2.65, and will vary depending on the metal precipitated and the chemicals used. As can be seen, a simple tank will not remove all of the suspended solids. The small particles must be brought together before they will settle from the water. This process is called flocculation.

There are two basic steps to flocculation. The particles all have the same charge on their surfaces, usually a negative charge. This charge is what keeps the particles separate. The first step is to neutralize this charge so that the particles can come into contact. Once the particles are in contact, they will not separate unless subjected to high shear forces.

Once the charge is neutralized, the second step is to agglomerate more and more particles together. The particles require gentle mixing so that they come into contact with other particles, but not with so much force that the contacting particles are broken apart. Floc is the precipitate that forms when the particles agglomerate. Flocculated solids should never be run through a centrifugal pump. The shear forces in the pump would easily tear apart the fragile floc.

The best way to increase the size of the particles depends upon the concentration of the suspended solids. The higher the concentration, the more contacts that will occur between the particles. At low concentrations, the particles are so far apart that the gentle mixing would have to continue for a very long period of time for all of the particles to contact the other particles.

One method to improve the efficiency of flocculation is to introduce coagulants. When the pH is raised with lime, both pH adjustment and coagulation occur. Coagulants enhance the formation of floc. Typical coagulants among the hydrolyzing metal ions are aluminum sulfate (alum), ferrous sulfate (copperas) and ferric chloride. Alum is used the most often with dosages ranging from 10 to 40 mg/l. Polymers constitute another group of coagulants. The polymers attach to different particles, bringing them together. One to five mg/l of these compounds is usually sufficient to increase the settling rate by two or three fold. Another advantage of using the polymers is that the solids capture is improved.

While coagulants improve flocculation, the main problem to over-come is the effect of low concentration on the frequency of particle contact. The equipment design must solve this problem. The type of clarifier used in the groundwater treatment system will depend upon the concentration of solids.

Settling Equipment

There are several types of settler designs that can be employed to remove suspended solids. Here we will review four designs:

1. Clarifier/thickener,
2. Flocculating clarifier,
3. Solids contact clarifier, and
4. Lamella.

These four designs are reviewed with the purpose of identifying the major components of each design. On short term groundwater treat-ment projects, the design engineer may not want to spend the money on a preconstructed settling tank. An existing tank or a portable tank can serve as a quiescent tank in which solids can settle. For successful settling, the velocity of the water must be reduced to a point where solids will settle by gravity. A rough guide regarding surface loading rates is that they should range from 0.2 to 0.7 gpm/sq. ft. of surface area for particle removal.

Clarifier/Thickener

A clarifier/thickener is the simplest of the four settler designs. A clarifier/thickener must perform several tasks. First, the water, with the suspended solids, must enter the tank without causing turbulence in the tank and reagitating already settled solids. Next, the water must be evenly distributed throughout the tank to make maximum use of the surface area of the clarifier. The supernatant must then be col-lected and removed from the tank. Finally, the solids must be thick-ened and removed from the tank.

Figure 5-5 depicts a standard "center feed" clarifier. The water enters into the center feed well. The feed well protects the contents of the clarifier from the energy contained in the incoming water. If the flow is to be introduced into the side of the tank, a plate or half-sphere (outside facing out) should be placed at the inlet pipe. This will serve the same energy dissipation function as does the feed well.

The keys to water distribution in the clarifier are the influent section and the effluent collection section. The influent must be

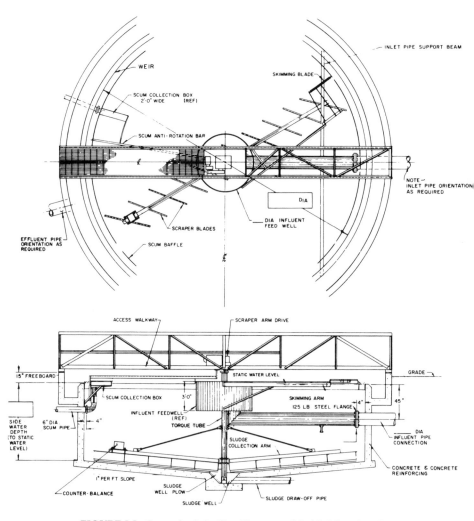

FIGURE 5-5. Center feed clarifier. (Courtesy of Smith & Loveless, Inc.)

introduced as specified above. The effluent must be collected over as large a surface area as possible in the tank. This is accomplished through the use of a saw-toothed weir on top of a trough. The maximum overflow rate should be limited to 13 gpm/ft of weir length. Lower levels are always better. The weir should be level throughout the tank to ensure proper distribution of the water in the tank. A flat weir can be used, but it is more difficult to adjust to the correct height throughout the clarifier. The troughs are connected to single pipe which exits from the clarifier.

The solids settle to the bottom of the tank. With flocculating solids, it is important to have sufficient depth for the floc to grow inside the clarifier. The depth also allows the solids to concentrate before they are removed from the tank. The less liquid that goes with the solids, the less material that has to be sent for disposal. The solids must be moved to one location for removal from the tank. The clarifier depicted in Figure 5-5 uses scrapers on the bottom of the tank to push the solids toward the center of the tank. The solids are collected in a sludge well, and exit the tank through a pipe. An alternative for scrapers is to slope the bottom of the tank at 60 degrees to a central point. This is called a hopper bottom clarifier, and is only viable for small designs.

Clarifiers should be designed at 0.2 to 0.7 gpm/sq. ft. of surface area. The maximum diameter of a portable clarifier is 12 ft. (wide load on a truck). Therefore, the maximum flow rate of a portable clarifier is about 80 gpm. Multiple units or a site erected clarifier will have to be employed for flows above this level.

Flocculating Clarifier

A flocculating clarifier has all of the same components of a regular clarifier. The only difference is that the influent well is expanded in size. Gentle mixing is performed in the influent well as the well is a flocculation zone inside the clarifier. A standard clarifier is good for solids of 1000 mg/l and above. Below that level, the clarifier looses efficiency. The flocculation section extends that efficient operation down to 500 mg/l. The specific concentration for both operations depends upon the settling rate of the solids. The flocculation section can be set up outside of the clarifier. The important design considerations are for gentle mixing and transfer of the water to the clarifier without any shear forces on the floc that has been formed.

Solids Contact Clarifier

The solids contact clarifier contacts already settled solids with incoming solids. The key to settling is large particles. The settled solids, of which a portion is recirculated, act as the core onto which the new solids attach. There are also more contacts between solids than would be provided with only a low concentration influent alone. Figure 5-6 shows a solids contact clarifier. Two advantages are realized with this design. First, lower concentrations of suspended solids can be intro-

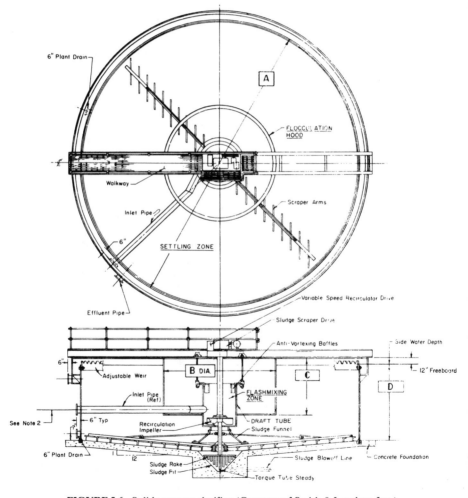

FIGURE 5-6. Solids contact clarifier. (Courtesy of Smith & Loveless, Inc.)

duced to the clarifier and removed efficiently. Second, the recycle of the solids makes more efficient use of the chemicals added to the water. Savings of 20-30% can be realized. The solids contact clarifier can be designed at up to 1.0 gpm/sq. ft. and influent solids as low as 100 mg/l.

Lamella Clarifier
The lamella clarifier has the highest flow rate per tank surface area of all of the clarifier designs. The lamella design is based on all of the basics of clarification, with one exception. It emphasizes the fact that once the solids hit the bottom of the tank, they are removed from the water. Instead of putting the bottom 10 feet from the top, the lamella puts the bottom two to four inches from the top.

Figure 5-7 illustrates a lamella clarifier. The water enters the sides of the plates. It is equally distributed to all plates. The water travels up the plates and the solids settle onto the plates. Once the solids have settled on the plates, they are removed from the water. The solids on

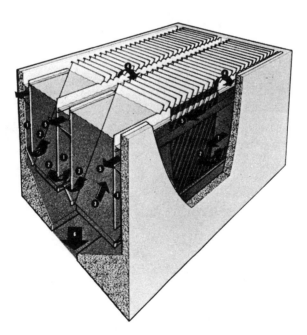

FIGURE 5-7. Lamella clarifier. (Courtesy of Graver Water.)

the plate continue down the plate as in a hopper bottom clarifier. The slope of the plates is from 45 to 60 degrees.

What makes the lamella design so powerful is that the theoretical settling area is the projected area of the plates, Figure 5-8. The projected area of all of the plates is additive. The resulting projected area can be ten times the liquid surface area of the tank. Therefore, up to ten times the liquid flow can be applied to the same size tank.

The solids that are removed with the plates fall off the plates and

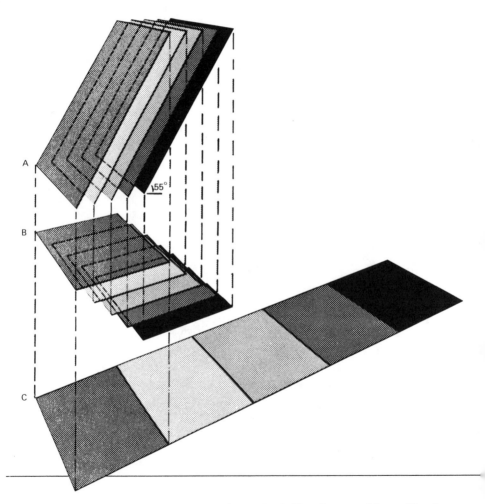

FIGURE 5-8. Theoretical settling area of a lamella clarifier. (Courtesy of Graver Water.)

enter the thickening section. The thickener can be of a hopper bottom design or it can have scrapers. The removal efficiency is the same as a standard clarifier. The design engineer uses the same laboratory data to design the lamella, but uses the projected area of the plates for the surface loading rate.

For flows that are small enough for a portable clarifier, 80 gpm, the lamella is probably not economical. When the tank is small, the cost of the plates makes the unit relatively expensive. However, for larger flows, the resulting savings in total size of the tank makes the cost of the plates economical. Even in large sizes, the lamella is a portable clarifier. The unit increases in length as more plates are added to increase the projected surface area, however, the 12 ft. wide load limit can be maintained for large units.

Filtration

All suspended solids are not typically removed by settling. Even the most efficient clarifier will leave 5 to 10 mg/l suspended solids in the water. This concentration is generally acceptable for reinjection, discharge to a receiving body of water or a POTW. However, when the water is to be used for drinking water or process water, even this low concentration is too high. The suspended solids must be removed by other technology, i.e., filtration. Filtration is also typically required as pretreatment for the implementation of such technologies as carbon adsorption reverse osmosis and other technologies requiring water with minimal amounts of suspended solids.

Low suspended solids must be filtered from the water. The simplest form of filtration is to pass the groundwater through a bed of sand. The suspended solids attach to the sand particles, and the water continues through the bed. It is very important for the design engineer to realize that a sand filter does not just strain the solids from the water. A sand filter removes suspended solids by several mechanisms including straining, adsorption, flocculation, and sedimentation.

The straining process occurs in the top few inches of a bed near the water/filter medium interface. Only the large particles are removed in this manner since they are too big to penetrate the voids between the medium. Some of the smaller particles will be removed deeper in the bed as particles agglomerate in the voids. However, it should be noted that straining is not the principal removal mechanism of a filter.

Adsorption (solid-solid contact) is the principal removal mechanism in filtration operations. The adsorption mechanism is similar to the way colloidal particles combine to form a floc. Adsorption is dependent upon the physical characteristics of the suspended solids and filter medium, and include filter grain size, floc size, adhesive characteristics, and the intensity of the surface charge on both the suspended solids and the filter medium.

Flocculation and sedimentation within the voids of the medium also occur but are of limited importance. With the tremendous amount of sand in the path of the water, the chances of contact are greatly increased. So, even with very low concentrations of suspended solids, the particles will come in contact with a sand particle, and be removed.

Polymer addition upstream of the filter may increase the removal efficiency of the filter. Polymers as filter aids can act as either coagulants or flocculants. Coagulants are more commonly used. Coagulants are adsorbed onto the filter medium surface and enhance the adsorption process. Typical coagulant feed rates when used as filter aids are from 0.25 to 1 ppm. Flocculants as filter aids strengthen the floc formed in the filter thereby improving shear stability. They are fed at dosages from 10 to 100 ppb. Over dosage can cause filter blinding. The removal efficiency of the filter can be improved from 50 to 90% by using these techniques.

As the filter bed becomes clogged and the pressure differential across the bed increases, it must be backwashed. Clean water is pumped through the sand bed opposite the flow of the wastewater at a fast rate. The sand bed is fluidized, and the turbulence in the bed breaks off the particles attached to the sand. The water flushes the suspended solids out of the filter. The sand is heavier and larger than the suspended solids and remains in the tank. When an increase in shear forces is required to remove the suspended solids, air is added to the sand bed to increase turbulence. An air scower system should be included in most filters to enhance backwashing capabilities used in groundwater treatment systems. The small extra capital expense is worth the potential problems that may result if it is not present.

A good example of sand filter operation occurred at a federal superfund site. A pilot plant was designed for 100 gpm and included an air stripper, followed by a sand filter, and a carbon adsorption unit. The sand filter was a modified pressure vessel without an air scour system for backwashing.

The treatment system was designed to treat several volatile organic compounds. The air stripper and the carbon adsorption units were chosen for organic removal. The sand filter was designed to protect the carbon adsorption bed from becoming fouled with suspended solids. The groundwater contained 4 to 7 mg/l iron in addition to the organic contaminants that the system was designed to remove.

The sand filter failed to protect the carbon bed under field operation. Two main problems caused this failure. First, the iron had not completely oxidized to the insoluble form before it reached the sand filter. The iron oxides precipitated out in the carbon adsorption bed. Second, the sand filter backwashing operation could not remove all of the iron solids from the sand bed.

The air stripper served as the main source of oxygen for the groundwater. This led to some iron oxide precipitation on the medium in the air stripper (See Chapter 3, Air Stripping section for a full review of this problem), and to large production or iron suspended solids from the air stripper. However, the air stripper only provided 4 to 7 minutes of reaction time for the iron to oxidize before it reached the sand filter. As discussed before in this chapter, this is less than the recommended 30 minutes reaction time for full oxidation. Therefore, some of the iron continued in the soluble form through the treatment system. But, since there was oxygen present in the water, the oxidation reaction continued. Therefore, some of the iron bypassed the sand filter in the soluble form, and then precipitated in the carbon adsorption unit.

The second problem came from backwashing the sand filter. Iron can be very difficult to remove from sand particles. This is especially true if some of the solid iron actually formed on the sand itself. The original design for the sand filter included only a water backwash for removal of the suspended solids. The water backwash operation was not able to provide the shear forces needed to dislodge the iron solids from the sand. The solids built up in the sand bed. Backwashing was required more and more often. Finally, "rat holes" occurred in the sand bed and the water short circuited through the bed, and the groundwater was not treated.

Both of these factors caused and carbon bed to fail because of suspended solids fouling. The pilot plant was modified to include an air scour in the sand filter, and the pilot test was completed. For the full scale design, preaeration with a retention time of 30 to 40 minutes was implemented to completely oxidize the iron. The suspended solids were then removed in a continuous backwashing sand filter, protecting the air stripper and carbon units.

As shown in Figure 5-9, a dual media filter consists of 1 ft. of anthracite and 1 ft. of sand. In a sand filter, there would be 2 ft. of sand instead of 1 ft. of sand and 1 ft. of anthracite. The water enters the top

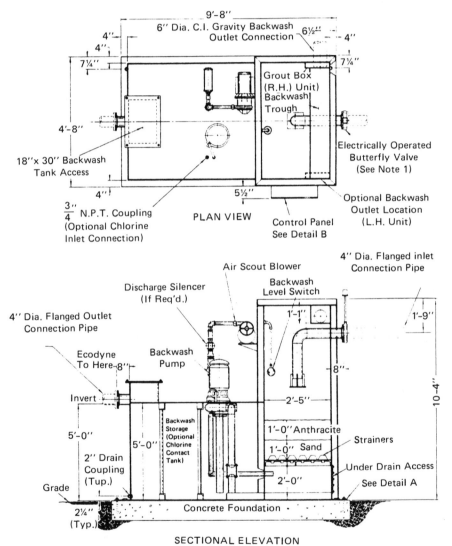

SECTIONAL ELEVATION

FIGURE 5-9. Dual-media gravity filter. (Courtesy of Smith and Loveless, Inc.)

and hits a splash plate. The splash plate ensures that the water falling onto the sand does not disturb the bed of sand. Sand filters can also be run in the upflow mode. There are advantages and disadvantages to this flow pattern, but most sand filters today are the downflow variety. Upflow continuous backwashing sand filters will be described later in this section. The distance from the top of the filter chamber to the top of the sand should provide enough room for a 50% expansion of the sand bed during backwashing.

When a filter bed is backwashed, the smallest particles end up on top of the bed. As solids are filtered out of the water, they build up in the sand. The smaller the sand particles, the faster the solids build-up. The solids will finally fill most of the void spaces in the sand bed. When this happens, the primary filtration mechanism switches from adsorption to straining. Straining of solids from the water takes a great deal of force, and the water builds up on top of the bed and the differential pressure increases. Gravity filters are normally set up with a water column above the bed to force the water through the sand. These columns can be up to 15 ft. above the sand, but are typically 7 to 10 ft. above with a minimum of 3 ft.

There are two main ways to extend the use of the sand bed. The first of these is to place a bed of coal above the sand bed, Figure 5-9. The coal particles are slightly larger than the sand, and they are lighter than the sand. When the beds are backwashed, the larger coal particles end up on top of the sand. This is referred to as a dual media filter. Suspended solids are removed by the coal before they ever get to the sand, thus extending the length of time between backwashing. Any time a filter is to be used after a biological system, or with a high concentration of suspended solids (50 to 100 mg/l), the dual media should be used.

The second way to extend the time between backwashes is to seal the filter tank and to use a pump to increase the pressure available for forcing the water through the filter bed. Pressure filters are good for high concentration suspended solids, 100 to 250 mg/l. The filter run must be long enough to produce sufficient water for backwash. The lower the percentage of processed water used for backwash, the better. A maximum of 10% of the processed water should be used for backwash, and the design should strive toward 2 to 5%. On the other hand, the filter should be backwashed every 24 to 48 hrs. This will ensure that the bed remains clean and that there is no build up of solids in the filter bed. The choice between sand, dual media and pressure should be made to address these system limitations.

Continuing down the filter, the next section is the sand support and backwash water distribution. There are two main ways to accomplish these functions. In Figure 5-9 the method is to use strainers on top of a support plate. The sand cannot get through the slots in the strainer. The backwash water and air are equally distributed to all of the strainers. The second way is to place a gravel bed under the sand and have distribution pipes inside the gravel bed. Once again, there are advantages and disadvantages to both designs. For the portable, groundwater market, the system with the strainers is probably the better system.

The final section of the filter is the backwash storage section. The water to be used for backwashing should be relatively clean of suspended solids. The water that has already been processed by the filter can be used for backwash. Backwash water should flow at 15 gpm/sq. ft. of filter surface area for 5 minutes. This water can be stored beside the tank and pumped at the necessary rate, or the water can be stored above the filter and flow back through the filter by gravity. Air, if desired, should flow at 3 to 8 scfm per sq. ft. of surface area for 5 minutes. A blower should be used to supply the air.

One final advantage of the sand and dual media filters is that their operation can be set up to be automatic. A pressure or time setting can be used to initiate the backwash cycle. No operator attention is necessary for the proper operation of these filters.

An alternative to the standard gravity and pressure filters is a continuous backwashing sand filter. There are two types of these filters currently available: a downflow and an upflow model. Both of them work on the same general principal whereby the filter medium (sand) is continuously cleaned by being recycled through an air lift sand washer. An upflow model by Parkson is presented in Figure 5-10. Such filters eliminate the need for backwash pumps, backwash storage, and valving associated with the backwash operation.

Filtration by Straining

Other types of filter technology are available for groundwater treatment. Sand filters rely upon adsorption, and avoid straining as a mechanism for suspended solids removal. However, several filter designs rely

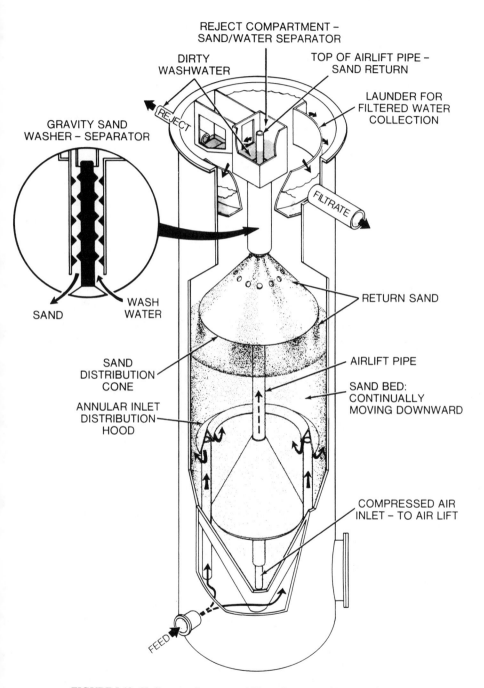

FIGURE 5-10. Upflow continuous sand filter. (Courtesy of Parkson Corp.)

upon straining. Bag filters, cartridge filters, and cloth filters all strain the suspended solids from the water.

Bag filters are basically a filter cloth in the shape of a bag. The bag is placed in a pressure vessel with a wire support for the bag. The water enters the center of the bag and the filter cloth strains the suspended solids from the water. The solids build up in the bag. Once the pressure drop is too high, the bag is removed and cleaned or replaced.

Bag filters have low filter surface to volume ratios, but a large solids storage area. This type of design is applicable to relatively large discrete solids. When the solids are small or if they are gelatinous (i.e., bacteria), the filter cloth with clog quickly. Large particles (i.e., sand) typically build up in the bag without large pressure drops.

Cartridge filters use the same straining mechanisms, but have a larger filtering surface area. The groundwater flows from outside in through the cartridge. The filter can be made up of several types of material, and can be designed for surface straining or depth removal. The best design for groundwater application is usually the wound type of depth cartridge. This design has a greater loading capacity and will remove small particles. Once again, if the pressure drop is too high, the cartridges should be cleaned or replaced. Surface cartridges can be cleaned, but depth filters usually have to be replaced.

Both the bag and cartridge filters are usually used to protect the next unit operation in the treatment system from solids fouling. Groundwater treatment does not normally require the low suspended solids concentration that these filters can achieve. The main reason why these units are employed for groundwater systems is low capital costs. Since groundwater flows are normally small, bag and cartridge filters can be a very economical way of removing suspended solids. Small sand filters, 10 gpm, with automatic air/water backwash will cost about $10,000. Bag or cartridge filters will cost about $1,000 for the same flow range. The bag and cartridge filters will have the added operating cost of cleaning or replacing the filter. The design engineer will have to perform a cost analysis to see which system is appropriate for a particular situation.

The cloth filter can also be employed to remove suspended solids. In this case, a filter cloth is placed across the groundwater flow. The cloth removes the suspended solids. These units are limited to low pressure drops across the filtering area, and are, therefore, limited in solids capacity. The main example of this technology as used for groundwater treatment is in conjunction with the diffused aeration systems discussed in the Air Stripping section of Chapter 3.

MISCELLANEOUS METHODS

Ion Exchange

The ion exchange properties of soil have been recognized since the 1850s. Since that time, there have been many improvements in the materials that can exchange an ion in the water for an ion on the solid phase exchange material. The largest use of ion exchange technology today is the use of synthetic resin beads for the softening of home potable water.

Ion exchange is basically the exchange of an ion with a high ion exchange selectivity for an ion with a lower selectivity. Any divalent ion will usually have a higher ion exchange selectivity than will a monovalent ion. Table 5-1 summarizes the selectivity for different ions on a variety of ion exchange resins. Calcium, which is divalent, will replace sodium, which is monovalent, at an exchange site on an ion exchange bed. This is the basis of water softening. The calcium ion, which increases water-hardness, exchanges with the sodium ion on the ion exchange resin. The calcium is removed from the water and the water has lost the ions that make it hard.

The resin is regenerated by passing a high concentration of sodium ions through the ion exchange bed. All reactions go in both directions. The calcium will exchange with the sodium, but at the same time the sodium will exchange with the calcium. The difference is that the rate of exchange for the calcium to replace the sodium is much faster than is the opposite reaction. However, a high concentration of sodium

TABLE 5-1   Ion Exchange Resin Selectivity[a]

Resin	Selectivity[b]
Strong acid	Li^+, H^+, Na^+, NH_4^+, K^+, Rb^+, Cs^+, Mg^{2+}, Zn^{2+}, Cu^{2+}, Ca^{2+}, Pb^{2+}
Weak acid	Na^+, K^+, Mg^{2+}, Ca^{2+}, Cu^{2+}, H^+
Strong base	F^-, OH^-, $H_2PO_4^-$, HCO_3^-, Cl^-, NO_2^-, HSO_3^-, CN^-, Br^-, NO_3^-, HSO_4^-, I^-
Weak base	F^-, Cl^-, Br^-, I^-, PO_4^{3-}, NO_3^-, CrO_4^{2-}, SO_4^{2-}, OH^-

[a]From Paterson, J. W. "Wastewater Treatment Technology." Ann Arbor, Mich.: Ann Arbor Science, 1978.
[b]Increasing selectivity left to right.

ions in the water, relative to the calcium ions, will drive the exchange in the opposite direction. The ion exchange resins can be regenerated with sodium for removing further hardness from the water.

All of the heavy metals present in water are in the divalent or trivalent state, with the exception of hexavalent chromium. A simple home sodium-ion-exchange-unit will remove all of these compounds. However, the process is expensive, and the regeneration brine, with the heavy metals, will still have to be disposed of off-site. These two problems severely limit the use of ion exchange for large quantities of heavy metals. The best use of ion exchange is for very low concentrations and for final treatment before potable use. Ion exchange can also be used to remove anions, negatively charged particles. Chlorides, nitrates, sulfates, etc., can be removed by anion exchange resins. Hexavalent chromium is, in fact, removed by anion resins.

All ions can be removed by ion exchange. Sodium ions can be removed by ion exchange resins using hydrogen. Hydrogen ions exchange with the sodium ions in the water. Sodium has a higher exchange potential than hydrogen. Combining anion exchange resins in the hydroxide form with cation exchange resins in the hydrogen form will remove all of the ions in the water. The remaining hydrogen and hydroxide ions combine to form water.

This process is used to make ultrapure water for high pressure boilers. The same process could be used to treat groundwater when the contaminant is dissolved solids. Once again, this is a very expensive process. Normally, an aquifer would be abandoned instead of being cleaned of dissolved solids. When treatment is necessary, reverse osmosis or distillation would be the preferred method.

Reverse Osmosis

Reverse osmosis (RO) separates a solute from a solution using a pressure gradient to force the solvent through a membrane. The selection of the membrane material, the configuration, and operating conditions are critical to obtaining the desired results. The most common membrane materials are cellulose acetate, polyamide, and thin film composites. However, other materials do exist and research is currently being performed to develop superior materials. The various configurations for the membranes include materials which are spiral-wound, hollow fiber, tubular, and plate and frame.

Specified operating parameters are the following:

1. Pressure, which is generally dictated by membrane material,
2. Recovery rate, which is the percentage of feed water converted to product water, and
3. Flux rate, which is the flow rate of water that passes through a unit area of membrane.

The flux rate of water through a membrane is proportional to the pressure differential across the membrane. The higher the pressure, the higher the flux rate for a given membrane. The flux rate also depends on the material thickness of the membrane and the temperature of the feed water. Flux rates should be specified conservatively to provide for long-term operation of the membrane.

RO can be used to remove most inorganic contaminants from groundwater. However, the wastewater must undergo extensive pretreatment prior to RO because RO is expensive to operate and is typically not used for general metals treatment. RO is used to recover metals in plating operations. Its use should be limited to nitrates, sulfates, total dissolved solids, and naturally occurring inorganics with groundwater.

RO is also not a preferred method for treating small concentrations in organic compounds. Low molecular weight organic compounds pass through the membrane at rates greater then do inorganics. Currently, research is being conducted to enhance the effectiveness of organic compound treatment with RO. Limited successes have been achieved with the high molecular weight organic compounds.

RO systems are readily available. They are expensive to run, due to the high pressures required, typically around 200 to 400 psig and the stringent pretreatment requirements. The pH generally must be maintained between 5.5 and 7.5 to minimize fouling. Suspended solids have to be removed to the maximum extent possible. Great care must also be taken to ensure that no precipitation occurs in the RO module.

RO was tested on groundwater at an RCRA site to remove sulfates and nitrates from an aquifer contaminated with VOC.[4] Effluent requirements became the controlling factor in the design selection. Nitrate concentrations had to meet drinking water standards of 10 mg/l while sulfate concentrations only had to reach 250 mg/l. Even with a total solids concentration of between 10,000 to 20,000 mg/l in the influent, the effluent nitrate requirements controlled the system set up. One of the problems with the operation of the system was that nitrate removal was supposed to be maximized by increasing the pH. However, at raised pH levels, precipitation occurred in the system and fouled the membranes. A pH of 6.5 was finally used as a compromise. Under these conditions, a two stage system was necessary to meet the nitrate requirements.

Electrodialysis

Electrodialysis is a combination of membrane technology and ion exchange technology. Electrodialysis uses ion exchangers in membrane form. The driving force across the membrane is provided by electric current. The ions are removed from the water and they pass through the membrane, attracted by the opposite electric charge on the other side of the membrane.

The advantages of the system are that the residence time controls the amount of dissolved solids removed, and that the system can be run continuously with no regeneration required. The disadvantage of the system is that the water must carry an electric current. The cleaner the water, the more resistance to the current, which increases the cost of operation.

Distillation

Distillation is the evaporation and condensation of the water stream. The inorganics do not evaporate with the water, and are left behind. The condensate is purified water. The process requires heating of the water to increase evaporation rates and cooling of the air stream to condense the water vapor. Volatile organics will evaporate and condense with the water.

The cost of heating and cooling the water can be very expensive. However, new technology in the area has resulted in the development of multiple effect distillation. Basically, this process uses the same energy several times in the process. Multiple effect distillation has had broad applications in water desalination projects in the Middle East. While the technology is readily available, the cost is still relatively high for groundwater treatment.

References
1. Steel, E. W. and McGhee, T. J. *Water Supply and Sewage* 5th Edition, McGraw-Hill, New York 1979.
2. Powell, S. T. *Water Conditioning for Industry.* New York, McGraw-Hill, 1954.
3. Patterson, J. W. *Industrial Wastewater Treatment Technology* 2nd Edition, Butterworth Publishers 1985. James W. Patterson, 1985.
4. Gregory, R. and Palmer, P. Nitrate removal from ground water utilizing reverse osmosis. In *Proceedings of the National Outdoor Conference on Aquifer Restoration, Ground Water Monitoring and Geophysical Methods,* May 1990, Las Vegas, Nevada.

6

Field Application of Design Methods

CASE HISTORY—ASSESSMENT AND INITIAL REMEDIATION OF A SMALL SITE CONTAINING HYDROCARBON PARAMETERS IN THE SUBSURFACE

John M. Wilson and Berny Ilgner

Geraghty & Miller was retained by a small oil company to conduct a site investigation at one of their facilities located in a rural area of a southern state (Figure 6-1). The investigation was initiated by complaints from the store owner of hydrocarbon vapors. Additionally, two offset landowners were lodging complaints about vapors emanating from a ditch line and about free product present on a farm pond. These complaints resulted in the local emergency management agency closing the store because of hazardous and explosive conditions at the site. The investigation was conducted under the scrutiny of the State Department of Environmental Management (DEM), under an administrative order. After several initial investigations by the DEM, Geraghty & Miller was contracted by the oil company to conduct the investigation. Activities at the site drew a high level of interest from local and state officials who monitored the progress. Site work was hampered by legal proceedings conducted by various individuals who were in litigation with each other. These included the downgradient offset landowners who were suing the store owners, who in turn were suing the owner of the tanks, who in turn, since the tanks had only recently been purchased, was seeking compensation from the previ-

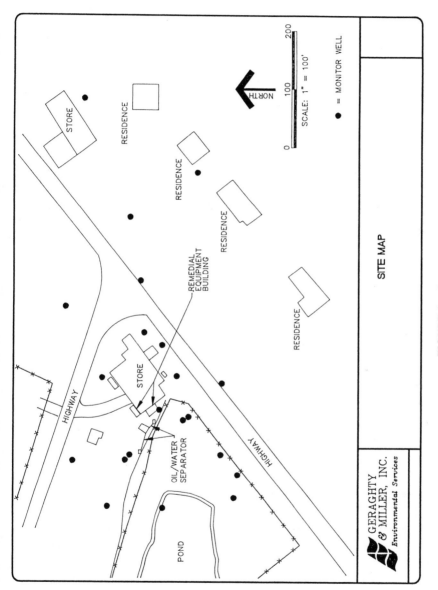

FIGURE 6-1. Site map.

SITE MAP

SCALE: 1" = 100'

● = MONITOR WELL

NORTH

STORE

RESIDENCE

RESIDENCE

RESIDENCE

RESIDENCE

RESIDENCE

REMEDIAL
EQUIPMENT
BUILDING

STORE

HIGHWAY

HIGHWAY

OIL/WATER
SEPARATOR

POND

GERAGHTY
& MILLER, INC.
Environmental Services

ous owner of the underground storage tanks (USTs). The legal activities indicated that in all likelihood the site data to be collected would eventually be used in settlements and in litigation.

Initially, Geraghty & Miller reviewed the administrative order to determine what conditions had to be met to comply with it. The first task included system tightness testing which was accomplished within the stringent schedule. These tests determined that all but one UST were leaking. One tank was leaking so profusely that it could not be tested. The tightness test results were submitted to the regulatory agency together with a preliminary work plan. The work plan included the following elements:

- Methodologies needed in order to accomplish the tasks. These included procedures to be used for sample collection of both soil and ground water; description of equipment to be used; decontamination procedures; specifications of well construction materials, including sizes, depths, thickness, material descriptions; and method of surface completion, and others.
- A complete sampling and analysis plan, including all quality assurance/quality control measures needed in order to obtain legally-defensible data.
- A complete health and safety plan to assure safety for site workers as well as nearby residents.

Following approval of the work plan, site work was undertaken to conduct three main tasks: (1) closure of all USTs; (2) site characterization and contamination assessment; and (3) stabilization activities (interim remedial measures).

Closure was accomplished with the removal of five small USTs. The water table was encountered at approximately 4 feet below grade, and hydrocarbon impacted soils and ground water were noted by field observations (Figure 6-2). After the sampling of ground water and soils, the pits were backfilled with clean sand packed into place.

As part of the site stabilization activities, a horizontal hydrocarbon vapor extraction system was installed. Trenches were dug along the tank pit boundaries between clean backfill and native material. Each trench was connected to a central trench which extended to the remedial building. Advanced drainage system (ADS) pipe with a filter sock was installed in each trench and plumbed to the remedial system. After these lines were laid, the trenches were backfilled to

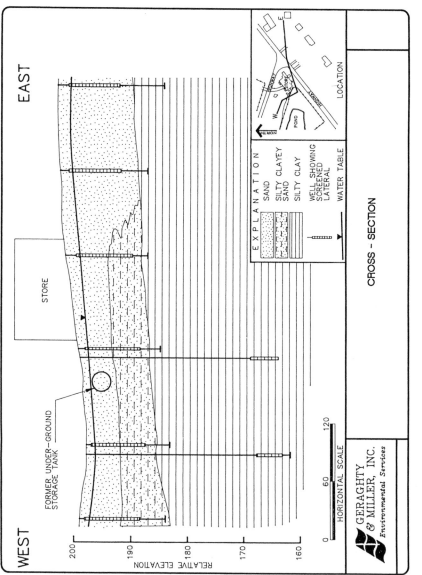

FIGURE 6-2. Cross-section.

251

grade. A recovery well was installed to collect liquid hydrocarbon. A skimmer pump was installed in the recovery well and plumbed to the remedial building where other equipment for the vapor extraction system was housed. The recovered hydrocarbon free product was retained in double-contained tanks with high tank shut-offs. As additional interim remedial measures, two small ponds were constructed in the ditch line (mentioned earlier in the complaint) to capture free-product migrating on the surface water (Figure 6-3). Hydrocarbon-absorbent pads were then floated on the ponds to absorb free-product. A discharge pipe placed through each dam and plumbed below the water surface allowed capture of the free-product. The ponds acted as in-ground oil/water separators. Additionally the contaminated soils excavated from the UST closure were stockpiled near the back of the site on and under plastic. A venting system consisting of ADS pipe was constructed within the mound and plumbed to the vapor extraction blower in the remedial building (Figure 6-4).

Concurrent with the tank removal and stabilization activities, characterization of the site and contamination assessment was underway. Included in the latter task was drilling, soil sampling, monitor-well installation, aquifer testing, and ground-water sampling. The purpose of this work was to obtain a general knowledge of the hydrogeological character of the site in the surrounding area in order to determine what monitoring and sampling was most appropriate. Five borings/wells were installed during this initial assessment. Findings of the preliminary investigation indicated that contamination encompassed the entire site and that additional assessment was needed. The geological findings at the site (Figure 6-2) included unconsolidated materials of sandy silty clays to a depth of approximately 15 feet where a gradual change to clay material at depth was observed. The water table across the site was uniformly projecting a westward flow and was 1 foot to 4 feet below grade.

From all of these site activities, a report was compiled that documented all results, procedures, and findings. Ground water at the site was determined to be contaminated in all wells screened in the surficial unit (15 feet) while a deep well (30 feet) was determined to be clean. Sampling of the farm pond indicated that the surface-water hydrocarbon concentrations were below detection limits. Although reports of free-product at the site were made together with detected vapors in the building, only minor free-product was detected in the UST pit, monitor wells, and recovery well.

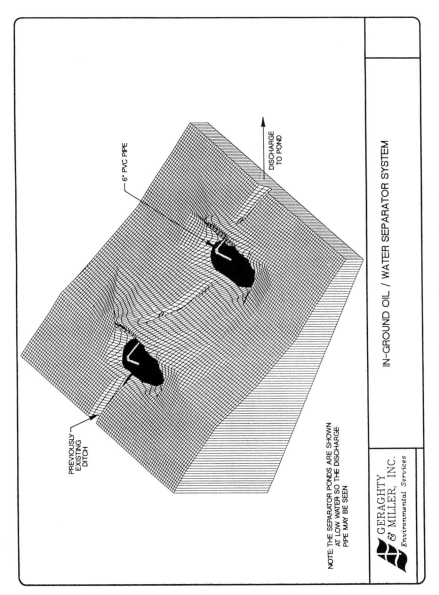

PREVIOUSLY
EXISTING
DITCH

6" PVC PIPE

DISCHARGE
TO POND

NOTE: THE SEPARATOR PONDS ARE SHOWN
AT LOW WATER SO THE DISCHARGE
PIPE MAY BE SEEN

GERAGHTY
& MILLER, INC.
Environmental Services

IN-GROUND OIL / WATER SEPARATOR SYSTEM

FIGURE 6-3. In-ground oil/water separator system.

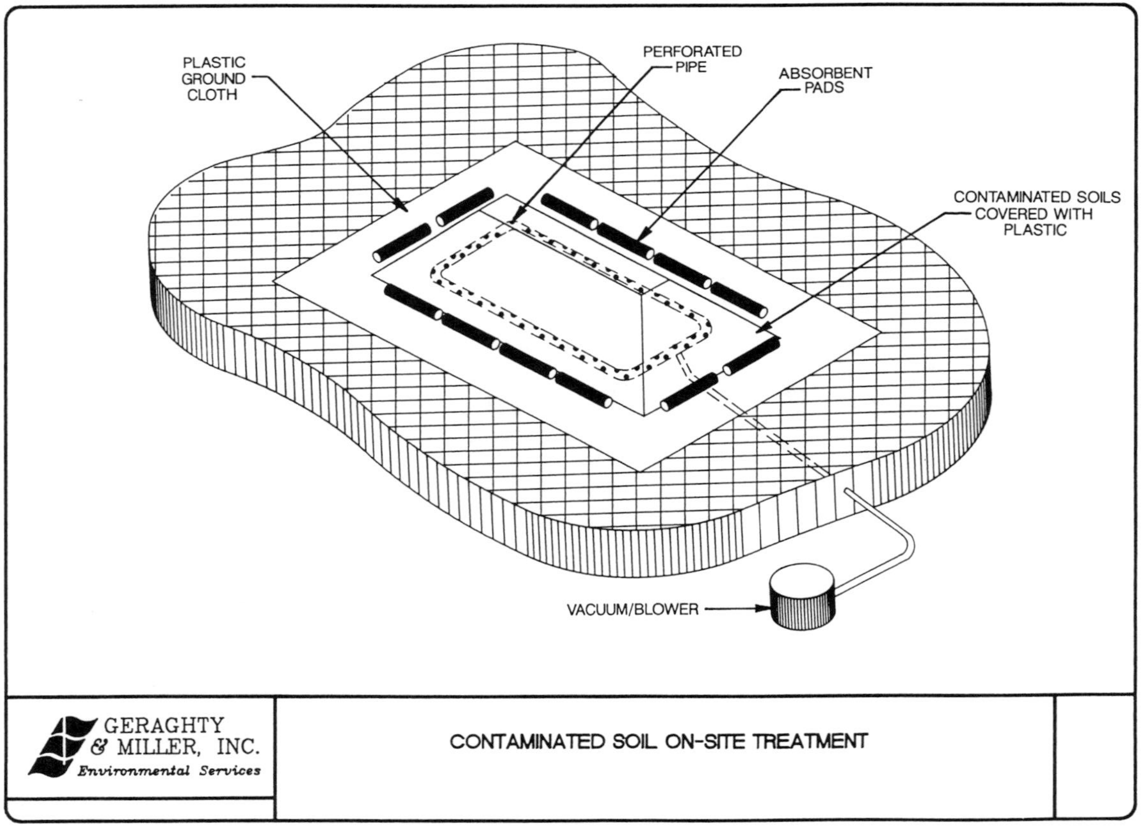

CONTAMINATED SOILS
COVERED WITH
PLASTIC

ABSORBENT
PADS

PERFORATED
PIPE

PLASTIC
GROUND
CLOTH

VACUUM/BLOWER

CONTAMINATED SOIL ON-SITE TREATMENT

GERAGHTY
& MILLER, INC.
Environmental Services

FIGURE 6-4. Contaminated soil on-site treatment.

254

Secondary and tertiary investigations were conducted in 11 more shallow wells and one more deep well. Information collected during several phases of the investigation included all data likely to be needed eventually by the engineers who were to prepare the corrective action plan. These data included:

- Direction of ground-water flow,
- Gradient of the water table which, together with porosity of the formation serving as the aquifer, was used to estimate the velocity of ground-water movement,
- Analyses of water included not only hydrocarbon components used to determine the vertical and horizontal extent of contamination, but those parameters usually needed for design of any water treatment system (i.e., hardness, total and dissolved iron, total and dissolved solids, and pH),
- Presence of man-made features that might impact design and installation of the remedial system (i.e., above- and below-ground utilities, elevations of all wells and sewers or natural drainages, buildings both onsite and offsite, and
- All known regulatory concerns and required permits for installation, operation, and maintenance of a remedial system.

Several water-supply wells were identified in the area and were routinely sampled together with the farm pond during these site investigations. These water supplies were free of hydrocarbons according to laboratory analyses. The extent of dissolved hydrocarbons was determined, and it was confirmed that the tanks at the store were in fact the source and that contamination was present over a large area. The plume extended from the site in a downgradient direction toward the farm pond and along the ditch line that extended westward from the store.

From the site findings, a corrective action plan was prepared to address and recover primarily dissolved hydrocarbons and phase and adsorbed hydrocarbons. Figure 6-5 shows the layout of the treatment system. Both adsorbed and vapor-phase hydrocarbon recovery was complicated by the shallow water table; therefore, the initial vapor extraction system installed around the tank pits was proposed for areas where unsaturated soils were thickest. In areas where the unsaturated soils were less than 2 feet thick, a sprinkler system with nutrients was proposed to help biodegrade hydrocarbon constituents.

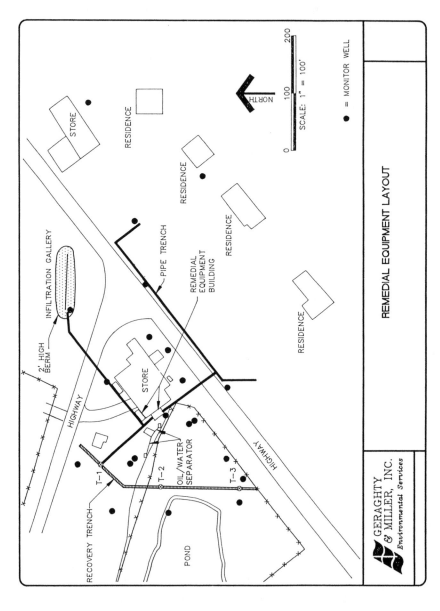

FIGURE 6-5. Remedial equipment layout.

The dissolved hydrocarbon recovery was the main focus for remediation. A pump and treat system was proposed to be the most effective remedial action for dissolved contaminants. Modeling was conducted to determine the most appropriate pump rates and well placements. The model indicated that two recovery wells, an infiltration gallery, and a trench with three sumps would be needed for the ground-water pump and treatment system. This scenario was discussed with engineering which then designed a corrective action plan to remediate the site. Figure 6-5 shows the layout of the remedial equipment.

Permits were sought for three different water discharges: (1) an NPDES permit for discharge to the ditch line, (2) an underground injection control (UIC) permit for an infiltration gallery located on the north portion of the site, and (3) a permit for the sprinkler system. All were needed to allow installation of the system. These permits along with permits allowing discharge of off-gases from the VES and air-stripping systems were sought while fabrication of the system and construction bids from subcontractors were assimilated. Operation and maintenance of the interim remedial systems were carried out until the full remedial system was installed.

AIR STRIPPING FOR A
MUNICIPAL GROUNDWATER SUPPLY*

Kevin Sullivan

Hydro Group, Inc., Linden, New Jersey

Air stripping has proven to be a cost-effective method of treating groundwater contaminated with volatile compounds. Its efficiency and economy have been proven at many hazardous waste sites and contaminated public water supplies, such as the case of Acton, Massachusetts.

The Water District of Acton (located outside Boston) traditionally had little treatment of their groundwater supply. However, in Decem-

---

*This section provided courtesy of Hydro Group, Inc.

ber 1978, the district lost 40% of its water supply when two wells were shut down after several organic chemicals—trichloroethylene, dichloroethylene, methylene chloride, and benzene—were detected. The two wells (Assabet No. 1 and No. 2) each have a yield of 0.5 mgd, and are located in the 375-acre Sinking Pond aquifer.

The district set out to discover the source of the contamination. A year-long hydrogeological study, including over 100 test wells, determined the probable cause of contamination to be the waste disposal practices of a nearby chemical plant. The main sources were a landfill and waste lagoons located 3000 ft away from the wells. The study also found an extensive plume (10,000 ppb total hydrocarbons) to be within 1000 ft of Assabet No. 2 (Figure 6-6).

After initiating conservation measures and securing an additional water supply from a neighboring town, the water district then began investigating possible treatment alternatives for the Assabet wells. Concurrently, the legal aspects of the contamination came to court in April 1980, when the U.S. EPA filed the first suit in Massachusetts under the Federal Resource Conservation and Recovery Act (RCRA). Later, the water district filed a separate $3 million suit against the chemical company. The federal suit was settled by a consent degree in which the company agreed to assist in cleaning up the aquifer.

The water district chose activated carbon as its first method of treatment to meet their self-imposed maximum level of 1 ppb of any single organic, and no more than 5 ppb overall. In June 1982, water from Assabet No. 1 was pumped through two GAC adsorbers, each containing 20,000 lb of GAC. The carbon systems reduced the organic levels from an average of 42 ppb to below the district's standard of 5 ppb.

However, the high cost of carbon recharges quickly became a problem. Approximately every five months, a complete recharge of all 40,000 lb of carbon was necessary, which translated into an operating cost of $37,000. The carbon alone created an operating cost of $0.30/1000 gal. Additionally, the influent concentrations continued to rise, which would further shorten the life of the GAC.

At this point, Acton decided to investigate packed tower air stripping as a pretreatment. Hydro Group's Environmental Products Division was called in to run a pilot test and install a full-scale air stripper.

In June 1983, an on-site pilot test was conducted, using a small, mobile packed tower (Figure 6-7). Seven different runs were performed in a single day, examining the effect of liquid loading rates,

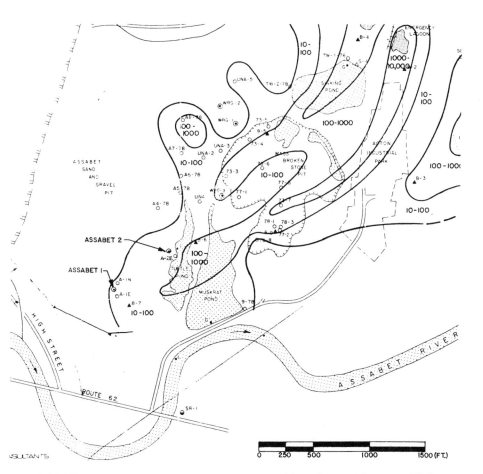

FIGURE 6-6. Contaminant concentration map, Acton, Massachusetts. *(Courtesy of Hydro Group, Inc.)*

air:water ratios, and packing height on the removal process. The rates and ratios for the test were based on previous tests conducted at similar sites; this allowed the scope of the test to be much more narrowly defined. Water loading rates varied between 10 and 30 gpm/ft^2, while air:water ratios ranged between 20:1 and 100:1. Samples were collected after 10 and 15 ft of packing and sent to a laboratory, along with raw water samples for analysis.

Once the samples were analyzed, percentage removal and mass transfer coefficients were determined for each of the runs. With these mass transfer coefficients, and the aid of some iterative computer

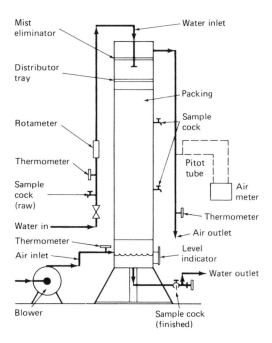

FIGURE 6-7. Mobile packed tower. *(Courtesy of Hydro Group, Inc.)*

programs, a variety of tower configurations could be modeled. The column was designed to handle 700 gpm and to provide 95% removal of the VOCs. Since the removal rate required was so "low" (towers with removal rates as high as 99.9% are in operation), the hydraulic loading on the column was increased to nearly 30 gpm/ft², allowing for a reduction in the diameter of the column and an associated cost savings. The air:water ratio was kept to a minimum in order to conserve on the blower energy requirements.

The final design of the tower was a 66-in.-diameter × 28 ft-overall-high unit, designed to hold a packed bed of 2-in. Tripacks 20 ft deep. The 5-hp blower will supply 4700 cfm at 3 in. water gauge static pressure to achieve a 50:1 air:water ratio.

Once the process sizing was completed, the structural design began. Wind, snow, and earthquake loads, in addition to the "dead" and "live" loads found in any structure, were considered. The weight of the packing and the retained water in the column under the operating conditions were also factored into the design (Figure 6-8). The col-

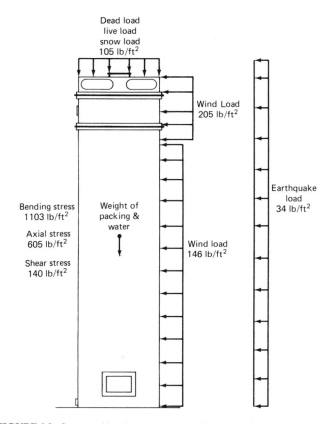

FIGURE 6-8. Structural loads on air stripper. *Courtesy of Hydro Group, Inc.)*

umn thicknesses were determined based on the structural properties of 6061-T6 aluminum.

The column was completely fabricated in Hydro Group's New Jersey manufacturing facility. Meanwhile, in Acton, the water district installed a sump to mount the air stripper on, as well as booster pumps to take the treated water from the stripper and deliver it through the carbon adsorption units into the distribution system.

The unit was delivered in February 1984. Installation of the unit took less than a day, using a small crane. When the unit was placed on line in April, the results were excellent. All organics were removed to less than 1 ppb. Removal percentages of 96-99% were recorded. These results are better than the predicted 95% due to the safety factors included in the full-scale design.

The capital cost of the air stripper was $31,000. Installation of the sump, a building to hold all of the electrical equipment, the pumps, and other miscellaneous equipment cost $109,000. Using 10% interest and a 20-year life, the yearly cost of capital equipment was $16,444. Electricity cost $0.06/kwh. The flow through the air stripper was 700 gpm. Total cost of air stripping the drinking water for Acton, Massachusetts, was $0.053/1000 gal.

Carbon life will be greatly extended by the air stripper at a fraction of the cost of carbon. Water district officials are very pleased with the installation, and plans are currently being developed to install another stripper at a recently contaminated well in the district.

LOW-CONCENTRATION ORGANIC REMOVAL FOR A DRINKING WATER*

Mark H. Stenzel

Calgon Carbon Corporation
Pittsburgh, Pennsylvania

Groundwater is a valuable resource for potable water. A recent survey indicated that 36% of municipal potable water supplies rely on groundwater, serving an estimated population of 75 million in 1980 (1). Many of these municipalities draw groundwater from large aquifers, extending well beyond their legal boundaries and providing water for other communities and other uses.

In many cases where groundwater sources have been found to be contaminated with dissolved organic compounds, nearby wells drawing from the same aquifer are not currently affected. This is due to the slow movement of groundwater and the associated plume of contamination. Users of the same aquifer need to be aware of the contamination, however, and should conduct periodic testing to determine whether the contaminants have reached their wells.

---

*This section provided courtesy of Calgon Carbon Corporation.

A typical case of aquifer contamination, and gradual spread of the problem, occurred in northern New Jersey in late 1980. New Jersey is a typical groundwater use area, as 47% of its public water supplies use this resource (1). In this case, Rockaway Township, a small community in Morris County, discovered that all three of its public wells were contaminated with trichloroethylene (TCE). In November 1980, the township installed a granular activated-carbon system to remove the contaminants.

Rockaway Borough, a nearby community of 7800 (Figure 6-9), became concerned about the quality of the drinking water it was drawing from the same aquifer and initiated a volatile organic scan program. Initially, only traces of tetrachloroethylene (PCE) were detected in one well, which was then shut down to prevent contamination of the entire supply. This testing confirmed earlier findings by the New Jersey Department of Environmental Protection, which was conducting tests in all towns near Rockaway Township using the same aquifer.

Within three months, the PCE level at this well had increased to 554 ppb, and the borough established a water use restriction program to cope with the loss of one of its wells. Borough officials realized that the contaminant plume would be moving into their well field and would eventually reach other operating wells.

"At this point . . . the Council and I were deeply concerned," recounted Borough Mayor Robert Johnson. "Closing down the one well only brought us temporary relief; we decided we had to find a way to remove the contaminant from our water" (2).

The mayor and Borough Administrator Walter Krich, Jr., immediately began to explore options for a permanent solution to treat the contaminated groundwater. Prior to installation of the treatment system (described later), however, contamination had spread to the other two wells, with detection of TCE as well as PCE.

On February 28, 1981, the mayor advised the community to discontinue use of tap water for drinking or cooking and announced that the borough's entire water system was to be shut down. Until a treatment system was in place, an emergency drinking water supply was made available using water trucks supplied by the National Guard, as shown in Figure 6-10, the County Department of Emergency Management, and a local dairy.

FIGURE 6-9. Rockaway Borough is located in Morris County in northern New Jersey. *(Courtesy of Calgon Carbon Corporation.)*

FIGURE 6-10. During the water emergency, the borough provided its residents with free, safe water from Civil Defense and National Guard tank trucks. *(Courtesy of Calgon Carbon Corporation.)*

NEW WELL AND TREATMENT OPTIONS

As already noted, the borough began to explore possible permanent treatment options even before the water emergency was declared.

Dormant municipal wells did not have enough output to supply the borough's 1.5 million gal/day average requirement. Borough land was previously tested for new wells, "but these areas are not sufficient sources of water," according to Krich. "The small excess capacity of nearby municipal systems, combined with 1980-81 drought restrictions on water use, preclude consideration of buying water on a permanent basis from another municipality," Krich said (2).

Aware of the success of carbon treatment at Rockaway Township, the borough contacted Calgon Carbon Corporation to evaluate the use of granular activated-carbon adsorption to treat the groundwater. Due to the low level of contamination, Calgon was only able to predict performance based upon carbon isotherm information (described in more detail in Chapter 3). A dynamic column study, if properly designed and run, would have taken many months to complete.

Supplier adsorption isotherm data, together with additional published data, was examined. Figure 6-11 shows a carbon adsorption isotherm for tetrachloroethylene published by Dobbs and Cohen of the EPA's Municipal Environmental Research Laboratory in Cincinnati, Ohio (3). This particular isotherm indicates a granular activated-carbon capacity of approximately 35 mg of tetrachloroethylene per gram of carbon at a tetrachloroethylene concentration of 550 ppb.

Based upon the favorable isotherm data for adsorption of tetrachloroethylene, installation of three single-stage carbon adsorbers was recommended. Each adsorber would contain 20,000 lb of granular activated carbon, and provide a superficial contact time of 15 min, treating one-third of the total flow. The 15-min contact time has been

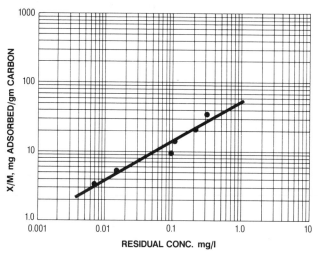

FIGURE 6-11. Carbon adsorption isotherm plot: tetrachloroethylene. *(Courtesy of Calgon Carbon Corporation.)*

proven to be sufficient for single-contaminant adsorption at low (ppb) concentrations.

The single-stage adsorber system, which was recommended to minimize capital expenditure, would not allow the maximum use of the granular carbon as predicted by the isotherm results. Calgon estimated that the single-stage capacity might be 60% of the theoretical capacity, or 21 mg/g.

At an influent concentration of 550 ppb, carbon usage was calculated as pounds of carbon per thousand gallons treated:

$$1000 \text{ gal} \times 0.55 \text{ ppm} \times \frac{8.34 \text{ lb PCE}}{\text{ppm } 10^6 \text{ gal}} \times \frac{\text{lb. carbon}}{0.021 \text{ lb. PCE}} = 0.22 \text{ lb carbon}$$

Based upon this usage rate, the recommended system, containing 60,000 lb of granular activated carbon, would require exchange of carbon after half a year of operation. It was stressed that this usage rate was valid for a flow of 1.5 million gal/day and a PCE contaminant level of 550 ppb. Once the concentration began to drop in the aquifer (due to flushing effects) substantially greater carbon life would be realized, lowering the overall treatment cost.

It was predicted that a two-stage system would come close to utilizing the full theoretical capacity of the carbon. The recommendation to use single-stage adsorbers was made, however, to minimize capital expense and in anticipation of either a lower flow or lower contaminant level which would extend carbon life.

Rockaway Borough accepted Calgon's proposal and adopted a municipal water utility bond for $700,000 to finance all costs associated with the water treatment system. This included a complete upgrade of the hydraulics of the water system and a building to house the adsorption system. The adsorption system shown in Figures 6-12 and 6-13—consisting of three adsorption vessels, 10 ft in diameter by 20 ft high; vessel internals (collection system); and 60,000 lb of Calgon Filtrasorb 300 granular activated carbon—was purchased from Calgon for $136,000. The borough also signed a three-year maintenance agreement with Calgon which would include future carbon deliveries as required.

The adsorption system components were shipped to the borough within six weeks, and the system began treating borough groundwater on June 25, 1981.

FIGURE 6-12. Within six weeks, Calgon Carbon Corporation delivered three vessels which would comprise the borough's carbon adsorption system. *(Courtesy of Calgon Carbon Corporation.)*

OPERATING RESULTS

With the initiation of carbon adsorption, the treated water showed no trace of TCE or PCE at the detection limit of 1 ppb and was declared safe for consumption, ending the water emergency.

Weekly testing has continued at the borough, to insure that the water is safe and to identify when the carbon system requires fresh

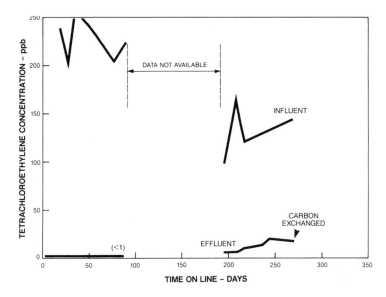

FIGURE 6-13. Operating results: June 1981 to April 1982. *(Courtesy of Calgon Carbon Corporation.)*

carbon. After flow at all well sites was reinitiated, PCE was detected at level up to 335 ppb, and TCE at 38 ppb, as influent to the carbon system.

The lower levels of PCE after normal water flow from the wells, and lower average water usage of 1.1 mgd, resulted in longer carbon life than predicted. The initial carbon bed remained on line until April 10, 1982, lasting 288 days. The second carbon bed then treated water until February 15, 1983, lasting 310 days.

Operating results for these two periods are shown in Figures 6-14 and 6-15. These figures illustrate the ability of granular activated carbon to treat a variable influent and provide a safe, consistent effluent, while allowing the user to make a decision when to exchange the carbon.

For the period shown in 6-14, the carbon system treated 362.8 million gal, resulting in a carbon usage rate of 0.165 lb of carbon/1000 gal. It should be noted that this usage was in a single-stage adsorption system, reducing PCE from 130 ppb PCE (avg.) to nondetectable levels. Fresh carbon produced nondetectable levels for about 10

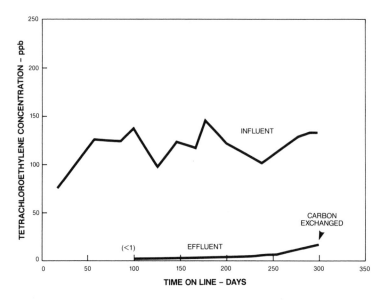

FIGURE 6-14. Operating results: April 1982 to March 1983. *(Courtesy of Calgon Carbon Corporation.)*

months; when PCE effluent reached 20 ppb the carbon would be replaced.

ECONOMICS

It is estimated that the borough spent $93,000 in the initial 12 months of water system operation for the carbon treatment system. This expenditure includes amortization of the adsorption system, the maintenance agreement (including additional carbon), and estimated utility costs. With this expenditure, the borough treated 410 million gal at an approximate cost of $0.227/1000 gal.

CONCLUSION

It can be seen from the Rockaway Borough experience that granular activated-carbon treatment of a contaminated groundwater source is a technically and economically feasible solution. The isotherm evaluation illustrates that the technology can be predicted, and the operat-

FIGURE 6-15. The carbon adsorption system is designed to treat 1.5 million gallons per day in three single-stage adsorbers. The system handles influent with tetrachloroethylene levels as high as 335 ppb. *(Courtesy of Calgon Carbon Corporation.)*

ing experience proves that it can accommodate variable flow and contaminant levels.

The attractiveness of carbon adsorption is best summed up by Mayor Robert Johnson: "How expensive was our solution: to finance the entire project of adsorbers and water system modernization, we raised our water rates $3.00 a month. Our minimum water bill went from $5.00 per quarter to $15. The average water bill in the Borough now is between $60 and $80 a year." This translates into approximately $0.76 per person per month.

"In the long run, we feel we have saved our community considerable money by walking the extra financial mile and installing a carbon adsorption system that makes our water pure, guaranteeing the good health of our citizenry," Mayor Johnson said (2).

BIOLOGICAL TREATMENT
OF A GROUNDWATER CONTAMINATED
WITH PHENOL

Evan K. Nyer

INTRODUCTION

A portion of the groundwater under a Gulf Coast hazardous waste site has been contaminated over the course of several years. From analysis of the groundwater and history of the site, several sources have contributed to the problem. These sources have been removed, leaving the cleanup of the groundwater the only remaining task to perform. However, this is the most difficult task, even though the plume is confined to the site boundary.

More than 30 wells have been constructed, and the extent and nature of the contamination plume has been well documented. A plan was developed that would take the water from several central wells, treat it, and then return the treated water through specially constructed recharge pits, forcing the plume back to the central wells. This section discusses the treatment of the groundwater to recharge quality.

The combined central wells have an average concentration of 15,000 mg/liter of dissolved solids and 1300 mg/liter of total organic carbon (TOC) with the main component being 400 mg/liter of phenol. For reinjection, this influent must be treated to background quality, which require a final effluent have less than 18 mg/liter of TOC. The groundwater is naturally brine, so the dissolved solids will not be removed. It was determined that the optimum pumping rate from the wells would be 23,000 gals/day. At this rate of removal and recharge, the design life of the treatment system was set at 10 years.

The process selection for the treatment system had to consider several important and unique problems connected with the treatment of a brine groundwater. Economics of various unit operations were used to determine the proper mix of technologies to reduce the concentration of organics down to the low levels required for recharge.

The most critical problem with the design of the treatment system was that the concentration of organics in the groundwater would decrease as the treated water was returned to the ground and forced back to the central wells.

Before the full-scale system was put into service, it was decided to run a large-scale pilot plant to insure that the assumptions made from the laboratory data were correct. The data from the first four months of operation are presented. Certain problems were encountered during startup, but overall, the system performed as expected.

PROCESS SELECTION

Several factors had to be considered in the process selection for the treatment system. The most important of these were economics and technical factors. A further consideration had to be given to the fact that the site preferred a relatively simple system that would not require full-time monitoring.

Laboratory data showed that the organics in the groundwater could removed by carbon adsorption or degraded by biological treatment. An economic comparison was then run on the two processes.

The cost of carbon adsorption is directly related to the pounds of organic removed. It takes between 5 and 200 lb of carbon for each pound or organic removed. For high concentrations of organics, as in this groundwater, the range is usually 5-20 lb of carbon per pound of organic. Using 23,000 gal/day and 1300 mg/liter of TOC, 249 lb/day of organics were needed to be removed from the groundwater. Using 10 lb/lb of organic for comparison, 2500 lb of carbon per day were needed. Assuming $0.75/lb of carbon, the operating cost is $1875.00/day. This number does not include capital costs and other operating costs such as personnel and electricity.

Biological treatment is usually considered as cost per 1000 gal of water treated. Standard numbers are based on relatively low concentrations of organics and high flow rates. For small flows (less than 100,000 gal/day) and high organic concentrations, a reasonable cost is $0.46/lb of organic treated. This number is high when compared to textbook numbers, but it is accurate for small-scale systems.

Using the same design numbers of 23,000 gal/day and 1300 mg/liter TOC, the operating cost is $115.00/day. This figure includes capital and electrical costs, but not personnel. This means that carbon will cost at least 16 times the cost of biological treatment.

These cost figures are based on the initial design specifications. The flow through the system will not change during the design life of the project. However, the organic concentration will decrease with time. The groundwater will be removed, treated, and returned to the aquifer. The treated recharge water will be used to force the plume back to the central wells. The treated water will mix with the plume as it forces it forward. This mixing will lower the organic concentration of the groundwater that is being removed. Figure 6-16 shows the expected decrease in organic concentration during the life of the project. These numbers will have to be confirmed after the treatment system is in full operation.

With the change in organic concentration, it was necessary to compare the costs of biological treatment and carbon adsorption over a range of organic concentrations. Figure 6-17 summarizes this comparison. As can be seen in Figure 6-17, the relative cost advantage of biological treatment over carbon adsorption does not change until approximately 150 mg/liter of organic concentration. The costs do not come close until less than 20 mg/liter. From Figure 6-18, the organic concentration is not expected to reach 150 mg/liter until six years after the start of the project.

From the economic analysis, it was determined that biological treatment of the groundwater was the preferred method. The only problem was that the treatment facility would have to consistently produce 18 mg/liter TOC for the recharge water. Based on experience

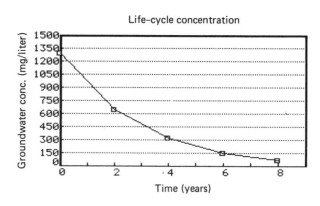

FIGURE 6-16. Expected decrease in organic concentration.

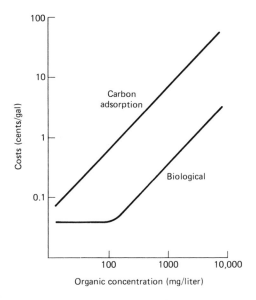

FIGURE 6-17. Comparison of costs of biological treatment and carbon adsorption.

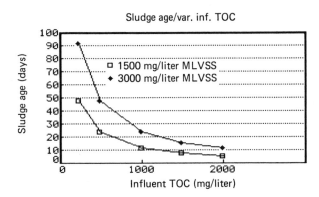

FIGURE 6-18. Effect of lower influent organic levels on sludge age.

and various literature on biological treatment, doubts were expressed that the biological system could consistently produce 18 mg/liter. The solution was to use biological treatment followed by carbon adsorption. This conclusion was arrived at without the use of a pilot plant study, because it was determined that the capital cost of a carbon system would be less than the cost of a pilot plant to test whether a carbon system was necessary.

LIFE-CYCLE DESIGN

There are several designs for biological treatment systems. The easiest to run are the fixed-film designs. However, the activated-sludge designs produce the best effluent; furthermore, several tanks were available on site could be used as part of the system, and an activated-sludge design could use the existing tanks.

Activated sludge was, therefore, preferable on a capital cost basis. However, there were two technical problems with using activated sludge. First, in order to maintain a consistent, low effluent organic level, the activated-sludge process requires close operator attention and daily analytical support. As mentioned before, these personnel requirements add substantial costs to the operations of the system. It was estimated that an activated-sludge design would require at least eight man-hours per day.

The second problem was that an activated-sludge system would not be able to adjust to the lower organic concentrations during the design life of the project. Figure 6-18 summarizes the effect of lower influent organic levels on the sludge age of the treatment system. This analysis assumes 23,000 gal/day influent flow, 40,000-gal aeration tank, and 0.25 lb solids/lb TOC removed yield coefficient.

The activated-sludge process relies on the ability of the bacteria to settle. A sludge age between 5 and 20 days is recommended to maintain a good settling sludge. As can be seen in Figure 6-18, the sludge age quickly goes out of that range as the influent concentration goes down. In the beginning of the project, a large aeration tank was required, and as the influent concentration goes down, the aeration basin must shrink to maintain the proper sludge age.

All of these considerations were combined and a final design developed. Figure 6-19 is the final design for the groundwater treatment system. The system includes a first-stage activated-sludge system, a second-stage fixed-film/activated-sludge system, a dual-media filter, and a carbon adsorption column. The following are the specifications on each section:

First-Stage Biological—The system consists of two 20,000-gal (15-ft diameter, 15-ft height) aeration basins, in series, and a hopper bottom clarifier. The first aeration basin has eight static tube aerators and a

1300 MG/L TOC

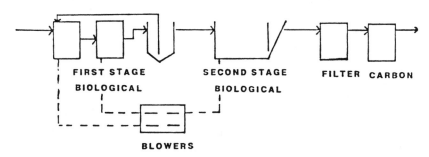

FIRST STAGE BIOLOGICAL   SECOND STAGE BIOLOGICAL   FILTER   CARBON

BLOWERS

LESS THAN 900 MG/L TOC

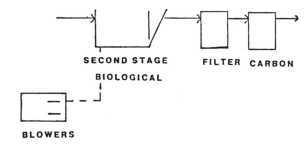

FIRST STAGE BIOLOGICAL   SECOND STAGE BIOLOGICAL   FILTER   CARBON

BLOWERS

LESS THAN 300 MG/L TOC

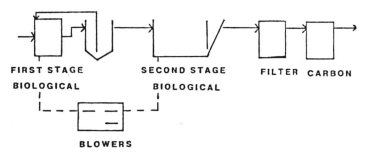

SECOND STAGE BIOLOGICAL   FILTER   CARBON

BLOWERS

LESS THAN 100 MG/L TOC

CARBON

FIGURE 6-19. Groundwater treatment schemes at various influent organic levels.

15-hp blower delivering 240 scfm of air. The second aeration basin has four static tube aerators and a 5-hp blower delivering 80 scfm of air. The hopper bottom clarifier has 97 ft^2 of surface area and returns sludge with an air, sludge ejector pump.

Second-Stage Biological—The second stage is the Fixed Activated Sludge Treatment System or FAST (registered trademark of Smith & Loveless, Inc.). The unit is 10 ft wide, 10 ft high, and 28 ft long. The system maintains the bacteria in the aeration zone by attachment to plastic media. The media are submerged in the water and the tank is completely mixed. The resulting system is an activated-sludge system, but it is no longer limited by the sludge age considerations. The system is also self-regulating and eliminates the need for operator attention.

Dual-Media Filter—The filter consists of 10 ft^2 filtering surface area, air/water backwash with 1 ft anthracite coal and 1 ft sand filter media.

Carbon Adsorption—The carbon columns consist of two carbon columns in series with 30 min residence time in each column.

The resulting system will be easy to maintain. There are no moving parts in the entire system, with the exception of the blowers and the pumps. The system will also require a minimal amount of operator attention. The first-stage biological system is designed to discharge up to 300 mg/liter of TOC. This flexibility will allow the operators to refine the operation a maximum of three days per week, and also will minimize the amount of analytical work required.

The FAST system is self-regulating and requires very little operator attention. The dual-media filter and the carbon adsorption systems are both fully automated. Manpower requirements are estimated at between 12 and 20 man-hours per week for the entire system.

The treatment system will also respond to the reduction in influent organic concentration. At 1300 mg/liter of TOC and above, the entire system will operate. When the influent TOC drops below 900 mg/liter, one of the aeration basins in the first-stage biological treatment system will be eliminated. When the influent reaches 300 mg/liter, the first-stage biological treatment system will go out of service. At 100 mg/liter, only the carbon will continue treating the groundwater. Figure 6-19 shows the treatment schemes at the various influent organic levels.

PILOT PLANT OPERATION

The full-scale plant design was based on preliminary laboratory tests and the economic and technical analysis presented here. There was a strong desire to have more concrete evidence that the biological treatment system would be able to degrade the organics in the brine groundwater. However, it was estimated that a continuous operating pilot plant would cost on the order of 25% of a full-scale system. It must be remembered that the full-scale system is only 23,000 gal/day.

The decision was to set up a pilot plant and have all of the material used on the pilot plant capable of being used on the full-scale system. The hopper bottom clarifier was purchased, and the second aeration tank of the first-stage biological system was set up with four static tube aerators and the 5-hp, 80-scfm blower. Both of these units would be used on the full-scale system. The pilot plant was estimated to be capable of treating 5000 gal/day.

On June 17, 1983, the pilot plant was started up. The initial seed for the reactor was provided by 8000 gal of waste-activated sludge from a nearby refinery and a commercial bacteria culture specific for hydrocarbons in salt water. The initial response of the bacteria was immediate. The dissolved oxygen in the aeration basin quickly went to less than 1 mg/liter. The influent flow was, at first, controlled by the level of dissolved oxygen in the tank.

The initial concentration of salt in the aeration tank was low. As brine groundwater was fed, the concentration increased. The salt had a definite effect on the bacteria. The mixed liquor suspended solids (MLSS) never increased past the initial concentration, and by day 14, started to decrease. Organic removal was on the order of 60% but the bacterial population would not increase.

At this point, the operator also switched methods for determining phosphorus and ammonia in the brine water. In both cases, the concentrations were found to be below desired levels. On day 26, 5000 gal of waste-activated sludge from a refinery that treated ballast

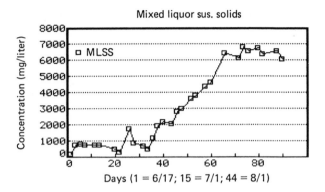

FIGURE 6-20. MLSS levels in the pilot plant.

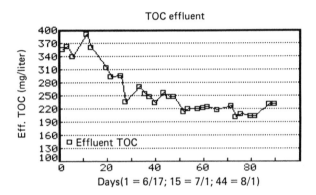

FIGURE 6-21. TOC concentration in the effluent for the pilot plant.

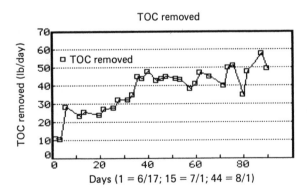

FIGURE 6-22. TOC removed by the pilot plant.

from oil tankers was added. These bacteria had grown in high concentrations of salt. At the same time, nutrients were added on a daily basis.

The bacterial population took about eight days to respond. After that, the MLSS continuously increased in concentration. On day 68, the MLSS reached 6000 mg/liter, and bacterial solids were wasted from the system. MLSS was maintained between 6000 and 7000 mg/liter. Figure 6-20 summarizes the MLSS levels in the pilot plant.

Effluent TOC responded to the increase in MLSS. Figure 6-21 summarizes the TOC concentration in the effluent for the pilot plant. On day 88, the influent flow increased to 15,000 gal/day. As can be seen in Figure 6-21, the effluent TOC increased with the added load. The cause of the increase was a faulty valve controlling the flow to the system.

Figure 6-22 summarizes the TOC removed by the pilot plant. In general, the TOC removed follows the same pattern as the MLSS and the effluent TOC. However, the same valve problem has made it difficult to control the flow to the pilot plant. The TOC removed has had a lot of variation due to the uncontrolled influent flow. The full-scale system will have a better control method for influent flow.

As can be seen by the above data, the organics in the brine groundwater are degradable by biological treatment. The entire treatment system was, therefore, installed.

FULL-SCALE OPERATION

The full-scale plant could not be put into operation at one time. The carbon unit could not be put into operation before the second-stage biological treatment system was operating. The dual-media filter was not required until the carbon unit was operational. The first step toward full-scale operation was to bring the second tank of the first-stage biological unit on line. The second step was to install the FAST unit.

The first-stage biological unit was able to have an immediate startup. This was due to established bacteria from the pilot operation. The FAST system took the more normal time of six weeks to start up. During this startup time, the effluent from the biological treatment systems was sent to an evaporation pond. Flow to the system was maintained at below full design flow to minimize the load on the pond.

Once the FAST system was performing, the dual-media filter and the carbon adsorption units were put into operation. Suspended solids from the FAST unit were very low, and the dual-media filter was put on a timed backwash cycle. The carbon was able to consistently remove the remaining TOC 18 mg/liter. The influent to the carbon system was a little higher than expected. The average influent to the carbon was 120 mg/liter. A project has now been started to seed the biological reactors with bacteria that can degrade some of the refractory organic compounds and reduce the influent TOC to the carbon units.

CONCLUSIONS

A portion of the groundwater under a Gulf Coast hazardous waste site has been contaminated with a variety of organic compounds. It was decided to pump the groundwater out of the ground, remove the organic contaminants, and recharge the cleaned water back into the ground. Initial influent concentration was 1300 mg/liter of TOC, flowing at 23,000 gal/day. Recharge concentration was set at the background concentration of the groundwater, 18 mg/liter of TOC. Over the life of the project, the influent concentrations will approach the background concentration.

A treatment scheme was developed based on laboratory tests, and technical and economic analysis. The final system included the following: a first-stage activated-sludge system, a second-stage fixed-film/activated-sludge system, a dual-media filter, and a carbon adsorption column. This system was economical and could easily be changed to reflect the changes in the influent concentration.

To insure that the organics in the brine groundwater could be degraded by bacteria, a pilot plant was set up. The pilot was designed so that all of the components would be used on the full-scale system. After overcoming salt inhibition and low nutrient concentrations, the pilot plant was able to consistently remove 70% of the TOC. The full-scale plant was started up in the first quarter of 1984, and it has steadily produced effluent TOC levels of less than 18 mg/liter.

IN-SITU BIOLOGICAL TREATMENT OF ISOPROPANOL, ACETONE, AND TETRAHYDROFURAN IN THE SOIL/GROUNDWATER ENVIRONMENT*

Paul E. Flathman
and
Gregory D. Githens

O. H. Materials Co.
Findlay, Ohio

INTRODUCTION

In-situ biological cleanup following spills of biodegradable hazardous organic compounds in the soil/groundwater environment can be a cost-effective technique when proper engineering controls are applied (4, 5, 6). Biodegradation of hazardous organic contaminants by microorganisms (7) can minimize liability by converting toxic reactants into harmless end products.

The cleanup of soil and groundwater containing isopropanol (IPA), acetone, and tetrahydrofuran (THF) by a combination of biological and physical techniques resulted in 90% removal of IPA and THF within three weeks. Acetone, an intermediate oxidation product of IPA metabolism, was removed by the end of the sixth week. The spill originated from several buried tanks which leaked contents into a shallow basin (12 ft) containing 100,000 ft^3 of sand and pea gravel.

The initial problem was to determine if biological cleanup could be utilized to remove isopropanol and acetone; 500 gal of THF were spilled by the client at an early stage of field operation. Acetone had not been spilled at the site but was suspected to be an incomplete oxidation product of isopropanol metabolisms (8).

---

*This section provided courtesy of O. H. Materials Co.

This case history describes

1. Bench-scale evaluation of the potential for biological cleanup in the spill site matrix
2. Field implementation
3. Removal rates of the contaminants at the spill site

Since the client is confidential, additional background information regarding this project cannot be presented. The underground recovery and treatment system, designed and developed by O. H. Materials Co., was used to effectively remove and treat the organic contaminants in the soil/groundwater environment.

O. H. Materials Co. has performed biological cleanups of spilled substances since 1978, when a railroad incident resulted in spillage of acrylonitrile. Subsequent biological environmental restoration projects have included additional acrylonitrile spills and other materials such as gasoline, crude oil, ethylene glycol, butylcellosolve, ethylacrylate, n-butylacrylate, methylene chloride, and various phenolics.

The case history demonstrates the practicality of biological detoxification of certain contaminants in the soil/groundwater environment. The data presented support O. H. Materials Co.'s earlier work and findings on groundwater restoration using biological techniques. Using the underground recovery and treatment system in combination with an activated-sludge biological treatment system, a recent project (6) also achieved greater than 50% reduction in the groundwater concentration of ethylene glycol within one week.

For this project, a recovery system was used to withdraw contaminated water from the ground for aboveground biological treatment. Effluent from the treatment system was reinjected into the subsurface environment, creating a closed-loop system. Biodegradation of the contaminants took place in the soil/groundwater environment as well as aboveground in the biological treatment system. The injection system was used to inoculate the underground environment with microbes capable of biodegrading the organic contaminants and to provide the inorganic nitrogen and phosphorus necessary to support microbial growth.

Cost effectiveness, minimal disturbance to existing operations, on-site destruction of spilled compounds, and permanence of solution are several of the advantages identified in this project for implementing biodegradation as a technique for spill cleanup and environmental restoration.

BIOFEASIBILITY EVALUATION

Prior to field operation, data were analyzed and a laboratory study was performed to determine whether a biodegradation program at the site was feasible. Results of analyses performed for soil and groundwater samples collected from the contaminated area and from a clean control area indicated that the value of all operational parameters were within favorable ranges for an enhanced biodegradation effort and only nitrogen and phosphorus needed to be augmented.

The biodegradation feasibility study was performed using static flask culture techniques (9, 10). Isopropanol-acetone-basal salts media (500 ml/liter Erlenmeyer flask) were inoculated within 12.5 ml (11) of a 1% (wt/vol) soil suspension and incubated at ambient temperature. The soil used for the inoculum was a representative sample collected from the spill site.

The medium employed (12) was prepared as follows:

Isopropanol	130 mg
Acetone	60 mg
KH_2PO_4	0.4 g
K_2HPO_4	1.6 g
NH_4NO_3	0.5 g
$MgSO_4 \times 7H_2O$	0.2 g
$CaCl_2 \times 2H_2O$	0.025 g
$FeCl_3 \times 6H_2O$	0.0025 g
per liter	

The pH of the medium was adjusted to 7.2 with $1N$ HCl or $1N$ NaOH prior to inoculation. An uninoculated control was prepared to quantify nonbiological loss of isopropanol and acetone.

At periodic intervals throughout the study, aliquots were removed from the reaction mixtures and analyzed for isopropanol and acetone and for pH and NH_3-N, NO_3-N, and PO_4-P concentrations to insure maintenance of a chemical environment favorable for bacterial growth. Isopropanol and acetone were quantified by headspace analysis on a Perkin-Elmer HS-6 unit (Perkin-Elmer Corporation, Instrument Division, Norwalk, Connecticut) in conjunction with a Tractor 560 gas chromatograph (Tracor, Inc., Instrument Group, Austin, Texas). The gas chromatograph was equipped with a 6-ft 1% SP-1000 on a 60/80

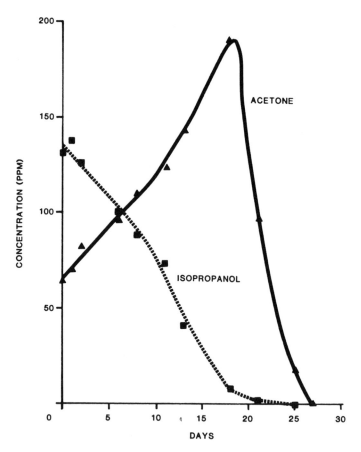

FIGURE 6-23. Feasibility data. Degradation of isopropanol and acetone in basal salts solution containing a soil inoculum.

mesh Carbopack B column with a flame ionization detector. Ammonium, nitrate, and ortho-phosphate were quantified spectrophotometrically by nesslerization, cadmium reduction, and ascorbic acid, respectively (Hach Chemical Company, Ames, Iowa).

Figure 6-23 indicates that isopropanol was biodegraded stoichiometrically in the reaction vessel to acetone. This reaction was biological since it did not take place in the uninoculated control. Daily biological loss during the first eight days of the feasibility study was 5.8 mg/liter, and nonbiological loss was only 0.39 mg/liter. Thus, the rate of biological loss for isopropanol was 15 times greater.

The concentration of acetone continued to increase until isopropanol was biodegraded to a trace amount and rapidly decreased once isopropanol was completely metabolized. The lack of a significant lag period indicated the presence of adapted indigenous microbes which could biodegrade the contaminants. This finding was significant because it indicated that in-situ biodegradation of isopropanol was already accurring and that OHM's management approach should be to increase the natural biodegradation rate. A biodegradation feasibility study was not performed for THF since it was not an original contaminant. THF, however, has been shown to be biodegradable by adapted microorganisms (13).

FIELD IMPLEMENTATION

Use of the underground recovery and treatment system for aquifer restoration has been previously described (5, 6, 14, 15, 16). All components of the system are easily assembled on site so that startup can take place within a matter of hours. With cleanup completed, those same components are easily disassembled and decontaminated prior to off-site removal to the next treatment location.

A schematic of the underground recovery and treatment system used for this project is presented in Figure 6-24. An activated-sludge system was the preferred method for aboveground biological treatment. In addition to providing efficient biological treatment, the activated-sludge system permitted wasting of adapted microorganisms for inoculation of the soil/groundwater environment through the injection system. With a flow rate of 6-10 gpm, the treatment system was

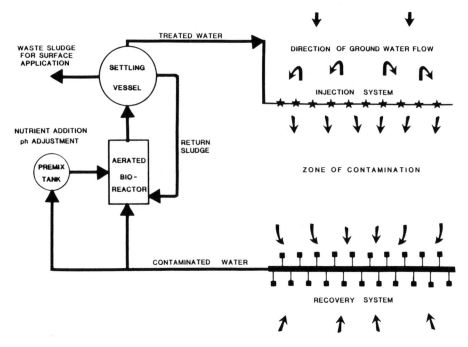

FIGURE 6-24. Schematic of the underground recovery and treatment system.

designed to provide a triple flush of the spill area within two months. The injection system was used to inoculate the underground environment with indigenous microbes capable of biodegrading the organic contaminants as well as to provide the inorganic nitrogen and phosphorus necessary to support microbial growth. The recovery system was used to withdraw contaminated water from the ground for aboveground treatment in an activated-sludge biological treatment system. Effluent from the treatment system was reinjected into the subsurface environment, creating a closed-loop system. Biodegradation of the spilled organic contaminants took place in the subsurface soil/ groundwater environment as well as aboveground in a biological treatment system.

Nitrogen and phosphorus nutrient additions and pH adjustments were made to the recovered groundwater from the recovery wells through the use of a premix tank to the bioreactor. From the bioreactor, the treated groundwater was pumped into a settling chamber from

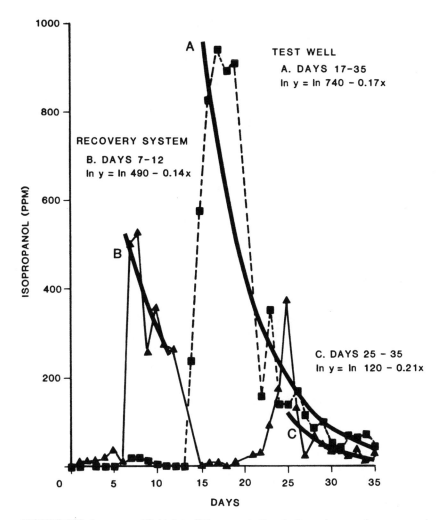

FIGURE 6-25. Isopropanol field data. IPA concentrations in the underground recovery and treatment system.

which a portion of the settled sludge was recycled into the bioreactor. Excess sludge was periodically wasted from the treatment system. The remainder of the treated groundwater was used for the injection system.

Results in Figures 6-25, 6-26, and 6-27 present respective IPA, acetone, and THF concentrations over a 35-day period for recovery

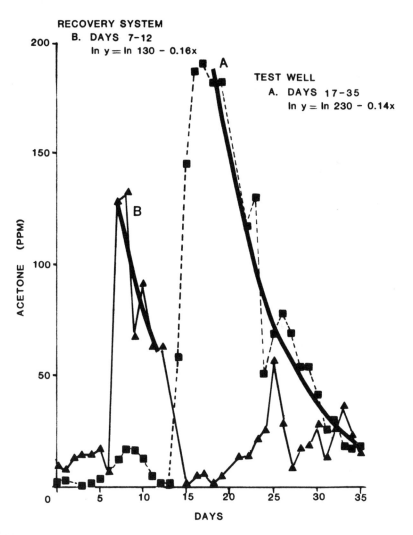

FIGURE 6-26. Acetone field data. Acetone concentrations in the underground recovery and treatment system.

system effluent and for a centrally located test well. Patterns of concentration change for IPA and acetone were very similar. The spike on day 15 for the test well reflects the flushing of contamination pockets following repositioning of the injection/recovery wells. Comparison of Figures 6-25 and 6-26 supports initial bench-scale findings of biological conversion of IPA to acetone. After day 15, acetone

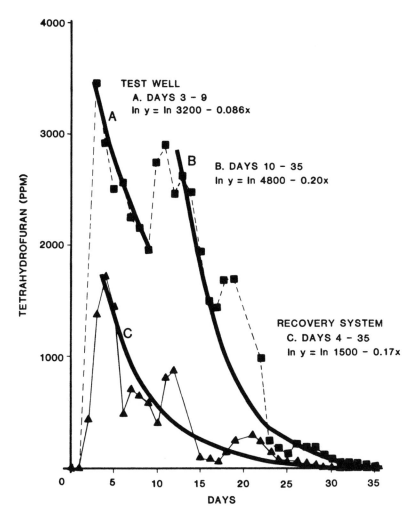

FIGURE 6-27. Tetrahydrofuran field data. THF concentrations in the underground recovery and treatment system.

concentration continued to increase until a low concentration of IPA was reached and then decreased very rapidly. By day 38, acetone concentration was less than 2 mg/liter and by day 44 was below the detection limit of 0.2 mg/liter.

Exponential decay curves were used to quantify removal rates of IPA, acetone, and THF from the groundwater environment. Holding other variables constant, the rates of decrease were assumed to be a function of contaminant concentration, that is,

$$\frac{dy}{dx} = -by$$

where:

y = level of contaminant remaining
x = time
b = rate constant

The curves generated in that fashion were fit to a first-order equation of the form

$$y = ae^{-bx}$$

where:

a = contaminant concentration at time zero

The first-order rate constant, $b,$ was determined by linear regression using least squares, and the first-order equation was converted to

$$\ln y = \ln a - bx$$

The calculated length of time for 50, 90, 95, and 99% removal of IPA, acetone, and THF from the test well and from the recovery system is presented in Table 6-1. Those rates reflect a combination of biological and physical removal. One line could not be fit to the data for each contaminant because physical changes in the injection/recovery system flushed out new pockets of contamination. Comparison of THF test well data following adjustment of the injection/recovery system on day 10 indicated that the calculated removal time was reduced by more than 50%.

TABLE 6-1   **Contaminant Removal Rates Using the Underground Recovery and Treatment System with Biological Techniques**

	Days				Sample Size n	Coefficient of Determination r^2
	50% Removal	90% Removal	95% Removal	99% Removal		
Isopropanol						
Test well						
Curve A	4	14	18	27[a]	17	0.86
Recovery system						
Curve B	5	16[a]	21[a]	33[a]	6	0.63
Curve C	3	11[a]	14[a]	22[a]	11	0.55
Acetone						
Test well						
Curve A	5	16	21[a]	33[a]	17	0.93
Recovery system						
Curve B	4	14[a]	19[a]	29[a]	6	0.73
Tetrahydrofuran						
Test well						
Curve A	8[a]	27[a]	35[a]	54[a]	7	0.94
Curve B	4	12	15	23	24	0.95
Recovery system						
Curve C	4	13	18	27	29	0.86

[a]Extrapolated values.

DISCUSSION AND CONCLUSION

The combination of physical and biological techniques was effective in removing those contaminants at a relatively rapid rate. The flexibility of the injection/recovery system in inoculating the site with biological and flushing media was a key determinant in the removal of scattered pockets of contamination.

For this project, two shallow basins of similar size and with similar contaminants were successfully treated using biological techniques. If it had been necessary to remove the contaminated soil from the two shallow basins treated at the site, the transportation and disposal (T&D) costs for 200,000 ft^3 would have been at least $550,000. The estimate for traditional T&D was based on a transportation cost of $3.00/mi/truck to a hazardous waste disposal facility 100 mi from the treatment site. Twenty cubic yards was the calculated capacity for each truck. The disposal cost for the contaminated soil was $60.00/yd^3.

By successfully decontaminating the soil and allowing it to remain on site, more than a five-fold cost savings was achieved and all future liability was substantially reduced. This estimate for cost savings, however, is conservative, since removal of soil from an area increases void volume for transportation and disposal by a factor of 1.2-1.3. Many times, hazardous materials cannot be shipped off site in bulk but must be packaged into drums. Additional chemical analysis are generally required to characterize the waste prior to off-site transportation and disposal. Inclusion of all these additional factors that must be considered for off-site disposal resulted in an even greater cost benefit to the client for in-situ cleanup.

Biodegradation as a method for spill cleanup and environmental restoration is considered a promising technology (5, 6, 17, 18). Land treatment techniques have been engineered and are accepted as an economical and environmentally sound means of destruction for many types of industrial wastes (5, 19, 20, 21). With regard to the cleanup of contaminated soil and groundwater, physical removal, by convention, has been a common method for remediation. However, biological techniques are now gaining increasing acceptance as a practical, cost-effective alternative for environmental restoration.

References
1. Pye, Veronica, I., Patrick, Ruth, and Quarles, John. "Groundwater Contamination in the United States." Philadelphia: University of Pennsylvania Press, 1983.
2. When a Water Supply Went Bad. *American City & County* Magazine, December 1981.
3. Dobbs, Richard A., and Cohen, Jesse M. Carbon Adsorption Isotherms for Toxic Organics. Cincinnati, Ohio: U.S. EPA Municipal Environmental Research Laboratory (EPA-600/8-80-023), April 1980.
4. Current Developments, Hazardous Waste. *Environmental Reporter,* May 6:11-12 (1983).
5. Flathman, P. E., Studabaker, W. C., Githens, G. D., and Muller, B. W. Biological Spill Cleanup. Proceedings of the Technical Seminar on Chemical Spills, Toronto, Ontario, Canada, October 25-27, 1983. Ottawa, Ontario, Canada: Technical Services Branch, Environmental Protection Service, Environment Canada, pp. 117-130.
6. Flathman, P. E., Quince, J. R., and Bottomley, L. S. Biological Treatment of Ethylene Glycol-Contaminated Groundwater at Naval Air Engineering Center, Lakehurst, New Jersey. Proceedings of the Fourth National Symposium and Exposition on Aquifer Restoration and Ground Water

Monitoring, Columbus, Ohio, May 23-25, 1984. Worthington, Ohio: National Water Well Association. In preparation.

7. Kobayashi, H., and Rittmann, B. E. Microbial Removal of Hazardous Organic Compounds. *Environ. Sci. Technol.,* 16:170A-183A (1982).

8. Stanier, R. Y., Doudoroff, M., and Adelberg, E. A. "The Microbial World," 3rd ed. Englewood Cliffs, N.J.: Prentice-Hall, 1970.

9. Barth, E. F., and Bunch, R. L. Biodegradation and Treatability of Specific Pollutants, EPA-600/9-79-034. Cincinnati, Ohio: Municipal Environmental Research Laboratory, U.S. EPA, 1979.

10. Tabak, H. H., Quave, S. A., Mashni, C. I., and Barth, E. E. Biodegradability Studies with Organic Priority Pollutant Compounds, *J. Water Pollut. Control Fed.,* 53:1503-1518 (1981).

11. Haller, H. D. Degradation of Mono-Substituted Benzoates and Phenols by Wastewater. *J. Water Pollut. Control Fed.,* 50:2771-2777 (1978).

12. Horvath, R. S., and Alexander, M. Cometabolism of *m*-Chlorobenzoate by an *Arthrobacter. Appl. Microbiol.,* 20:254-258 (1970).

13. E. I. du Pont de Nemours & Company, Inc. Tetrahydrofuran—Properties, Uses, Storage & Handling. Wilmington, Del: Publication no. E-62465, 1984, p. 28.

14. Ohneck, R. J., and Gardner, G. L. Restoration of an Aquifer Contaminated by an Accidental Spill of Organic Chemicals. *Ground Water Monitoring Review,* 2(4):50-53 (1982).

15. Quince, J. R., and Gardner, G. L. Recovery and Treatment of Contaminated Groundwater: Part I. *Ground Water Monitoring Review,* 2(3):18-22 (1982).

16. Quince, J. R., and Gardner, G. L. Recovery and Treatment of Contaminated Groundwater: Part II. *Ground Water Monitoring Review,* 2(4):18-25 (1982).

17. Dobbs, D., and Walton, G. C. Biodegradation of Hazardous Materials in Spill Situations. Paper presented at 1980 National Conference on Control of Hazardous Material Spills. Louisville, Ky., May 13-15, 1980.

18. Thibault, G. T., and Elliott, N. W. Biological Detoxification of Hazardous Organic Chemical Spills. Paper presented at 1980 National Conference on Control of Hazardous Material Spills. Louisville, Ky., May 13-15, 1980.

19. Loehr, R. C., Jewell, W. J., Novak, J. D., Clarkson, W. W., and Friedman, G. S. "Land Application of Wastes" (2 vols.). New York: Van Nostrand Reinhold, 1979.

20. Parr, J. F., Marsh, P. B., and Kla, J. M., "Land Treatment of Hazardous Wastes." Park Ridge, N.J.: Noyes Data Corporation, 1983.

21. Vernick, A. S., and Walker, E. C., eds. "Handbook of Wastewater Treatment Processes." New York: Marcel Dekker, 1981.

Index